Summer 2006

No 124

national **STATiSTiCS**

Population Trends

In this issue

palgrave macmillan

0525665

About the Office for National Statistics

The Office for National Statistics (ONS) is the Government Agency responsible for compiling, analysing and disseminating many of the United Kingdom's economic, social and demographic statistics, including the retail prices index, trade figures and labour market data, as well as the periodic census of the population and health statistics. It is also the agency that administers the statutory registration of births, marriages and deaths in England and Wales. The Director of ONS is also the National Statistician and the Registrar General for England and Wales.

A National Statistics publication

National Statistics are produced to high professional standards set out in the National Statistics Code of Practice. They undergo regular quality assurance reviews to ensure that they meet customer needs. They are produced free from any political influence.

About *Health Statistics Quarterly* and *Population Trends*

Health Statistics Quarterly and *Population Trends* are journals of the Office for National Statistics. Each is published four times a year in February, May, August and November and March, June, September and December, respectively. In addition to bringing together articles on a wide range of population and health topics, *Health Statistics Quarterly* and *Population Trends* contain regular series of tables on a wide range of subjects for which ONS is responsible, including the most recently available statistics.

Subscription

Annual subscription, including postage, is £100; single issues are £27.50.

Online

Health Statistics Quarterly and *Population Trends* can be viewed or downloaded as Adobe Acrobat PDF files from the National Statistics website www.statistics.gov.uk/products/p6725.asp *(Health Statistics Quarterly)* or www.statistics.gov.uk/products/p6303.asp *(Population Trends)*.

Editorial board

Joy Dobbs (editor)
Peter Goldblatt (editor)
Angela Dale
Judith Jones
Azeem Majeed
Jil Matheson
Ian R Scott

Contributions

Articles: 5,000 words max.

Dates for submissions

Title \ Issue	Spring	Summer	Autumn	Winter
Health Statistics Quarterly	by 11 Sept	by 11 Dec	by 22 Mar	by 21 June
Population Trends	by 23 Oct	by 2 Feb	by 4 May	by 26 July

Please send to:

Ian Thurman, executive secretary
Population Trends
Office for National Statistics
Zone D3/06
1 Drummond Gate
London SW1V 2QQ
Tel: 020 7533 5125
E-mail: ian. thurman@ons.gsi.gov.uk

Contact points at ONS

People with enquiries about the statistics published regularly in *Health Statistics Quarterly* and *Population Trends* can contact the following enquiry points.

Topic enquiries

Abortions: 020 7972 5537 (Department of Health)
    E-mail: abortion.statistics@dh.gsi.gov.uk
Births: 01329 813758
    E-mail: vsob@ons.gsi.gov.uk
Conceptions: 01329 813758
    E-mail: vsob@ons.gsi.gov.uk
Expectation of life: 020 7533 5222
    E-mail: lifetables@ons.gsi.gov.uk
Marriages and divorces: 01329 813758
    E-mail: vsob@ons.gsi.gov.uk
Migration: 01329 813872/813255
Mortality: 01329 813758
    E-mail: vsob@ons.gsi.gov.uk
Population estimates: 01329 813318
    E-mail: pop.info@ons.gsi.gov.uk
Population projections:
    National – 020 7533 5222
    E-mail: natpopproj@ons.gsi.gov.uk
    Subnational – 01329 813474/813865

General enquiries

National Statistics Customer Contact Centre
Room 1015 Government Buildings
Cardiff Road
Newport NP10 8XG
Tel: 0845 601 3034
E-mail: info@statistics.gsi.gov.uk
Website: www.statistics.gov.uk

ISBN 0-230-00319-2

ISSN 0307-4463

in brief

International migration 2004

Net in-migration rises

On 20 April 2006, the Office for National Statistics (ONS) published its annual reference volume *International migration* 2004, series MN no. 31. An international migrant is someone who changes his or her country of usual residence for a period of at least a year, so that the country of destination effectively becomes the country of usual residence.

In 2004, an estimated 223,000 more people migrated to the UK than migrated abroad. This is 72,000 higher than 2003 and is the highest net in-migration since the present method to estimate Total International Migration (TIM) began in 1991. All comparative statements about TIM are made solely with reference to the period 1991–2004.

This rise in net in-migration was mainly due to the number of people arriving to live in the UK for at least a year increasing from 513,000 in 2003 to a record 582,000 in 2004. The number of people leaving the UK to live elsewhere in 2004 was similar to levels seen in the last two years.

However, the number of British citizens migrating abroad continued to increase leading to a record net out-migration estimate of 120,000. There was also a record estimate of 342,000 for the net in-migration of non-British citizens.

Net in-migration of European Union (EU) citizens was 74,000 in 2004. Of this, 48,000 were citizens of the ten accession (A10) countries that joined the European Union in 2004. Net migration for citizens of A10 countries is set within the context of accession to the EU occurring in May 2004. Therefore, migration outflows will not feed through until future years. Fuller consideration of accession migration beyond 2004 is given as part of the National Population Projections: www.gad.gov. uk/Population/2004/methodology/migrassa8. htm

Migrants are not necessarily citizens of their country of last/next residence. For example, a British citizen migrating to the EU becomes an EU resident but remains a British citizen.

In 2004 there was a net in-migration to the UK of 41,000 from residents of A10 countries. These ten accession countries moved from 'Other' and New Commonwealth to EU grouping in 2004. For the EU15 countries, the recent trend of net out-migration continued in 2004 (24,000). Combining these flows gives a net in-migration from the EU25 countries of 17,000.

Figure 1	Total International Migration to/from the UK, 1991–2004

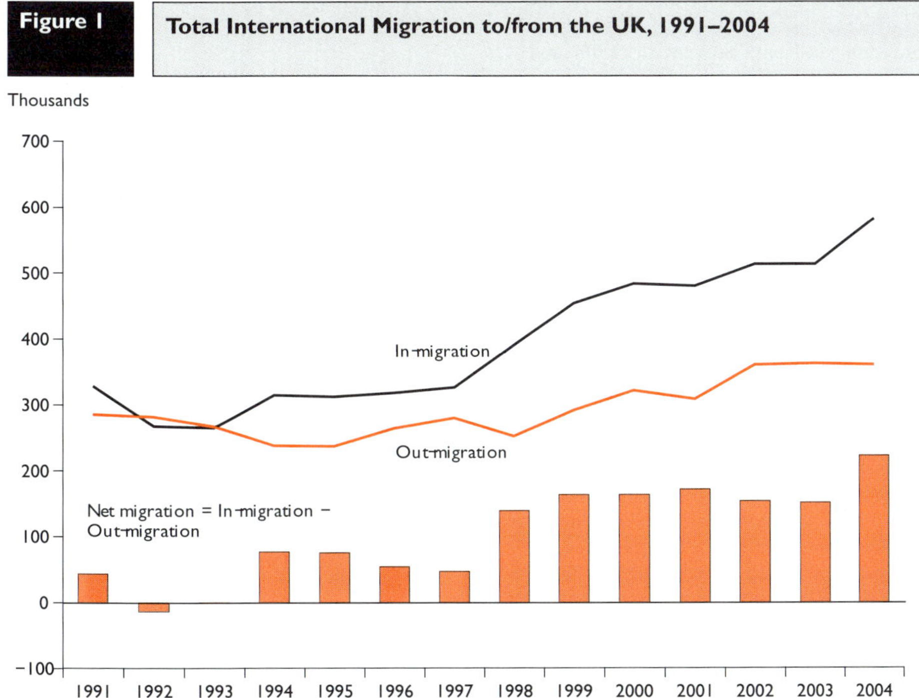

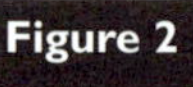

Figure 2   **Net migration by country of last or next residence, 1995–2004**

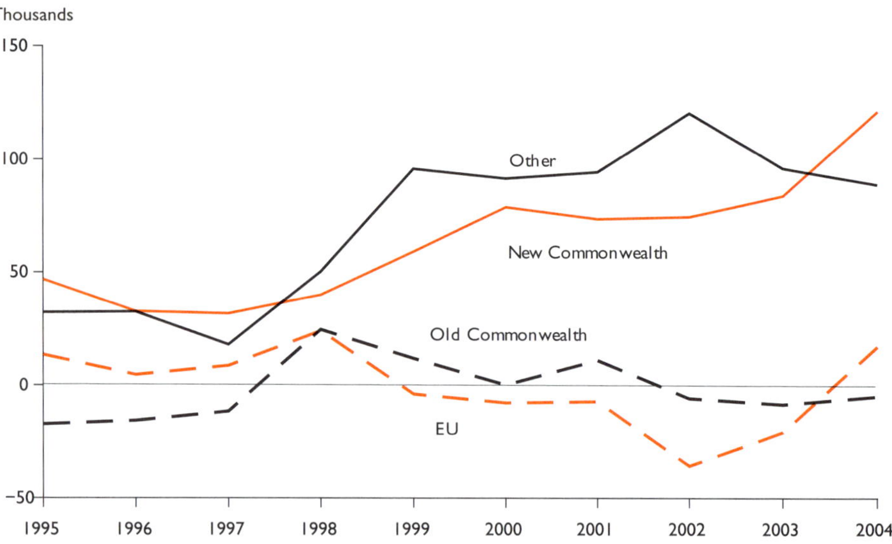

Net in-migration of New Commonwealth residents increased by 45 per cent between 2003 and 2004. Within this group, net in-migration from Bangladesh, India and Sri Lanka rose from 38,000 in 2003 to 54,000 in 2004. For Pakistan the proportional increase was even higher rising from 9,000 in 2003 to over 25,000 in 2004.

Although in-migration has increased over the decade, migrants are intending to stay for shorter periods. In 1995, 42 per cent intended to stay for more than four years compared with 34 per cent in 2004. Those intending to stay between one to two years increased from 36 per cent in 1995 to 50 per cent in 2004.

A link to *International migration* MN31 can be found at www.statistics.gov.uk/StatBase/ Product.asp?vlnk=507

UK electoral statistics, 2005

On 23 February 2006 ONS published *UK electoral statistics*, 2005. The total number of electors and attainers were published for both local government and parliamentary electors. The number of local government electors in the UK grew by 294,318 (0.7 per cent) to 44,944,109 and the number of parliamentary electors grew by 223,172 (0.5 per cent) to 44,403,415. There were increases in the numbers registered to vote in both the local government and parliamentary elections in all of the constituent countries of the UK.

The increases in the numbers registered to vote were due partly to the re-registration of those previously removed from the electoral canvass as a result of non-response for at least two years ('list cleaning'). Other reasons identified by the Electoral Registration Officers were a more extensive canvass to non-responders and various other initiatives including use of the media and poster campaigns.

The current Electoral Register came into effect on 1 December 2005 and represented the number of people who would be entitled to vote if an election had been held on 1 December 2005. It was based on a qualifying date of 15 October 2005.

There are 646 parliamentary constituencies in the UK. The two largest constituencies were the Isle of Wight and Daventry with 107,790 and 89,197 people registered to vote in parliamentary elections respectively. The two smallest were Na h-Eileanan an lar and Meirionnydd Nant Conwy with 21,404 and 32,830 respectively.

For England, Wales and Scotland the residence qualification required a person to be normally living at the address on the qualifying date, even if temporarily absent. People with more than one place of residence, such as students, may therefore be included on more than one register but are only entitled to vote in one constituency in a general election.

European\local government electors are those who are entitled to vote in European and local government elections and meet the residence qualification. These include European Union citizens but exclude overseas electors. Parliamentary electors are those who are entitled to vote in parliamentary elections for Westminster and meet the residence qualification. These include overseas electors but exclude Peers and European Union citizens. Overseas electors are not resident in the UK, but must previously have been registered in the UK and included in the electoral register (unless they were too young to register). They are registered in the same parliamentary constituency as before they went abroad. Attainers are also included in both these figures. Attainers are those who will reach the age of eighteen years during the currency of the register, and therefore will be entitled to vote at an election on or after their eighteenth birthday.

Electoral data are collated by ONS in England and Wales, and by the General Register Office for Scotland (GROS) and the Electoral Office for Northern Ireland (EONI).

The electoral data for 2005 and previous years, together with further information on electoral statistics, can be found from the electoral statistics home page on the National Statistics website by following the appropriate links from www.statistics.gov.uk/StatBase/Product. asp?vlnk=319&Pos=&ColRank=1&Rank=272

Mid-2005 population estimates, Scotland

On 27 April 2006, the General Register Office for Scotland published its *Mid-2005 population estimates, Scotland*. The key points of this report are:

- The estimated population of Scotland on 30 June 2005 was 5,094,800, a rise of 16,400 on the previous year and an increase of 30,600 since mid-2001.

- Compared with the previous year there were more births (+1.3 per cent) and fewer deaths (−1.8 per cent). Despite this, deaths exceeded the number of births by about 2,300.

- Over the year there was a net gain from migration of around 19,000, including a net gain of 12,500 people from the rest of the UK and a net gain of 7,300 people from overseas (including asylum seekers).

- Around 57,300 people came to Scotland from England, Wales and Northern Ireland and around 44,800 left Scotland to go in the opposite direction. The net inflow of around 12,500 is lower than the previous year's 15,500 net inflow reflecting less people coming to Scotland and more people leaving than the previous year.

- The lower overseas net gain of 7,300 comes from an inflow of around 35,400 (including asylum seekers) and an outflow of 28,100. Note that short-term migrants are not included in the estimate as an international migrant is defined as someone who changes their country of residence for 12 months or more.

- Since mid-2001 and the 2001 Census, Scotland's population has increased by 0.6 per cent (+30,600) from 5.06 million to 5.09 million. However, over the last ten years, Scotland's population has decreased by 8,890 (–0.2 per cent): 5.10 million to 5.09 million, due to an excess of deaths over births because net migration fluctuated around zero for most of the period.

- Amongst the Council areas, Aberdeenshire, Highland and Falkirk had the largest percentage increases over the year at 1.1 per cent, closely followed by Edinburgh City at 0.9 per cent. East Dunbartonshire and West Dunbartonshire experienced the largest percentage decreases at 0.6 per cent, followed by Midlothian and Aberdeen City both with percentage decreases of 0.5 per cent.

- Of Council areas which had been decreasing: Dundee City's population increased for the first time since mid-1995, and both Eilean Siar and Glasgow City increased for two years in a row.

- Of the NHS Board areas, Highland (+1.1 per cent), Forth Valley (+0.9 per cent), Lothian (+0.6 per cent) and Fife (+0.6 per cent) had the largest percentage increases. The only two NHS Board areas with percentage decreases were Argyll & Clyde (–0.4 per cent) and Ayrshire and Arran (–0.2 per cent).

- Population density is 65 persons per square kilometre and ranges from eight persons per square kilometre in Highland Council area to 3,298 persons per square kilometre in Glasgow City Council area.

Further details may be found at www.gro-scotland.gov.uk/statistics/library/mid-2005-population-estimates/index.html

Joint ESRC/ONS/ BSPS public policy seminar series on the implications of demographic change

Demographic changes in society have strong implications for Government policy as well as for what sorts of official statistics are needed so that sound decisions can be made. The Economic and Social Research Council (ESRC) in conjunction with ONS and the British Society for Population Studies (BSPS) have organised a series of three seminars on the implications of demographic change. They are part of ESRC's 'Mapping the public policy landscape' seminar series. The seminars, which are for invited attendees only, have the aim of engaging policy departments and academic experts to discuss the key issues with those who provide the statistics. Each of the seminars is accompanied by a brochure which is publicly available.

The first seminar on the topic of changing family and household structures and complex living arrangements took place on 18 May 2006 and information from it including the brochure is available at www.esrcsocietytoday. ac.uk/ESRCInfoCentre/about/CI/events/ seminarseries/index.aspx?ComponentId= 5938&SourcePageId=6066. Information about the second seminar, on population ageing (30 June), will appear at the same address shortly. Information about the final seminar on migration (21 July) can be found at www. esrcsocietytoday.ac.uk/ESRCInfoCentre/about/ CI/events/esrcseminar/index.aspx?Component ld=5048&SourcePageId=6063. For hard copies of the brochures or further information about the seminars please contact Amanda Williams at the ESRC on 01793 413126; e-mail: amanda. williams@esrc.ac.uk.

Corrections

2004-based national population projections for the UK and constituent countries

An article detailing the assumptions and results of these projections appeared in *Population Trends* 123. There was an error in Table 6 of that article, which shows the change in projected population by certain age groups compared with the interim 2003-based projections. The figures and percentages for the Under 16 and 16 to 29 age groups were shown incorrectly for all years. A corrected version of the table appears below.

The reference volume: National Population Projections 2004-based series PP2 No. 25 describing in detail the methodology, assumptions and results of these projections is scheduled to be published in July 2006.

Revision of stillbirths in England and Wales in 2004

It has recently come to light that some register offices in England and Wales failed to notify ONS, in line with regulations, of some stillbirths that occurred in 2004. The result is that the perinatal mortality figures (numbers and rates) for 2004 given in reference tables 2.1 and 6.2 in *Population Trends* (and in *Health Statistics Quarterly*) are undercounts. In addition, the stillbirth and perinatal mortality figures for some areas presented in the Report on Infant and perinatal mortality, 2004: health areas, which was published in *Health Statistics Quarterly* 27, autumn 2005 edition, are undercounts.

Similarly, the stillbirth and perinatal mortality figures for 2004 given in other ONS publications are also undercounts. These publications include *Mortality statistics: Childhood, infant and perinatal* 2004 (series DH3 no. 37) published on 28 March 2006, *Key Population and Vital Statistics* 2004 (series VS no. 31, PP1 no. 27), published on 27 April 2006, and *Birth statistics* 2004 (series FM1 no. 33) published on 15 December 2005.

Table 6   **Change in projected population by age compared with the interim 2003–based projections**

United Kingdom

Thousands/percentages

Age group	2004		2011		2021		2031	
	Thousands	Per cent	Thousands	Per cent	Thousands	Per cent	Thousands	Per cent
Under 16	6	0.1	112	1.1	250	2.4	240	2.3
16–29	59	0.5	264	2.2	232	2.0	384	3.4
30–44	–2	–0.0	142	1.1	416	3.4	432	3.4
45–59	–9	–0.1	–15	–0.1	–7	–0.1	229	1.9
60–74	–6	–0.1	–21	–0.2	–54	–0.5	–92	–0.8
75 and over	0	0.0	8	0.2	55	0.9	120	1.6
All ages	48	0.1	490	0.8	892	1.4	1,313	2.0

Work is currently underway to ensure all stillbirths that occurred in both 2004 and 2005 are correctly identified for inclusion in published figures. ONS aims to publish revised stillbirth and perinatal mortality figures for 2004, with the new figures for 2005 in *Health Statistics Quarterly* 31 in August 2006.

Recent Publications

Health Statistics Quarterly 30 *(Palgrave Macmillan, £27.50, May, ISBN 0-230-00315-X)*
International migration 2004 (series MN no. 31) *(April, available on the National Statistics website at www.statistics.gov.uk/StatBase/Product.asp?vlnk=507)*
Key Population and Vital Statistics 2004 (series VS no. 31, PP1 no. 27) *(Palgrave Macmillan, £40, April, ISBN 1-4039-9385-8)*
Mortality statistics: injury and poisoning 2004 (series DH4 no. 29) *(June, available on the National Statistics website at www.statistics.gov.uk/statbase/product.asp?vlnk=621)*

Office for National Statistics Spring Departmental Report 2006 *(TSO, £8, May, ISBN 0-10-168382-0) (Available from TSO 0870 600 5522 or at www.tso.co.uk/bookshop).*
Regional Trends 2006 (no. 39) *(Palgrave Macmillan, £41, May, ISBN 1-4039-9071-9)*

All of the above Palgrave Macmillan titles can be ordered on 01256 302611 or online at www.palgrave.com/ons. All publications listed can be downloaded free of charge from the National Statistics website.

Demographic indicators

England and Wales

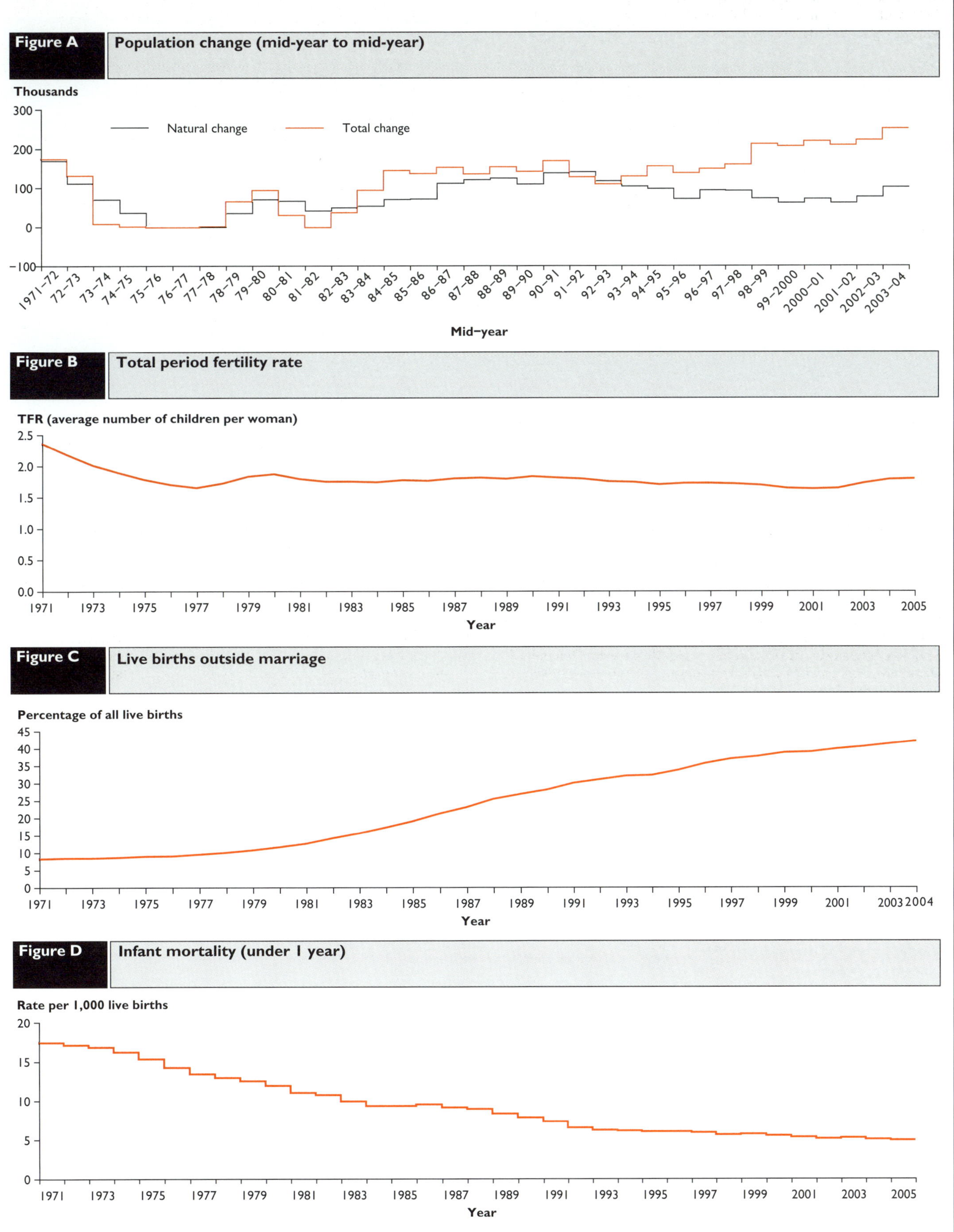

Estimates of the population by ethnic group for areas within England

Pete Large and Kanak Ghosh
National Statistics Centre for Demography
Office for National Statistics

This article describes, and provides some initial analysis of, the experimental population estimates by ethnic group for areas within England published by the Office for National Statistics (ONS) in January 2006.

The article considers growth and the population structure of each of the ethnic groups identified in the 2001 Census; subnational patterns of change; population turnover; and measures of diversity and segregation, and also provides a comparison of the estimates and corresponding sample-based estimates from the Labour Force Survey.

INTRODUCTION

Previous articles in *Population Trends*[1,2] have provided estimates of the sizes of different ethnic groups in Great Britain and its constituent areas. Those estimates have been based on the results of the Labour Force Survey and, though reliant on good quality data collected by interviewers, are subject to substantial sampling errors when estimating small groups or changes over time. In January 2006, the Office for National Statistics (ONS) published new, experimental, population estimates for England by ethnic group based on a cohort component methodology. This methodology, which was designed to avoid the difficulties in relying on sample survey-based estimates, was described in a previous article in *Population Trends*.[3] This article describes the estimates themselves, covering the extent and causes of population change in each ethnic group; differences in population structure; subnational patterns of development; and measures of 'ethnic diversity' and segregation of ethnic groups. A final section compares the population estimates by ethnic group with other sources.

METHODOLOGY

A full explanation of the methodology underlying the experimental population estimates by ethnic group is provided in a previous article in *Population Trends*.[3] In summary, the population for each ethnic group at mid-year is found by taking the previous estimate, ageing on one year, adding births, subtracting deaths and adjusting for migration. Differences in demographic rates, such as fertility rates and propensity to migrate, for each ethnic group are estimated using data from the 2001 Census. The base population is taken from the 2001 Census but adjustments made in the Mid-Year Estimate (MYE) series are also incorporated, as explained in the previous article[3], to ensure that the estimates by ethnic group are consistent with the MYEs for each year.

Box one

Definition of Ethnic Group

The complexities of defining and describing ethnic group are discussed in *Ethnic group statistics*.[4] Key points are:

- ethnic group is self-assigned – that is, chosen by the respondent from a list of categories (including an 'other' option)
- the classification used in National Statistics is the 16 way classification adopted in the 2001 Census (see, for example, Table 1)
- a person's ethnic group can change over time
- description of ethnic group can change in different contexts. Reliance on the census data in the modelling process has the *de facto* effect that the estimates will accord with the context of the census – in particular, this will reflect any effect due to proxy responses by the form-filler on behalf of another household member

It is an error to assume that a population is homogenous within an ethnic group. This error is compounded when groups are combined. For example, the term non-'White British' is used in this article when referring to people who belong to any ethnic group other than White British, but it is not asserted that the non-'White British' population can be treated as being a homogenous demographic group.

DATA QUALITY

The Population Estimates by Ethnic Group released in January 2006 are experimental statistics. This means that they have not yet been shown to meet the quality criteria for National Statistics, but are being published to involve users in the development of the methodology and to help build quality at an early stage (more information on experimental statistics and National Statistics is provided in the National Statistics Code of Practice: Protocol on Data Presentation, Dissemination and Pricing[5]).

The acknowledged limitations of the methodology must be borne in mind when interpreting the estimates. In particular, the methodology is based on reliance on 2001 Census data for parameter estimation. Thus, while the absolute level of, say, births in a year is taken from the registration data used in the Mid-Year Population Estimates, the estimation of the differences in fertility between ethnic groups is based on 2001 Census data, which provides the ethnic dimension not provided in the registration data. This method is robust to changes in the size of the population of an ethnic group but cannot track changes in differentials in the demographic rates. Further sources of potential error include:

- estimation of the starting population using the census, and subsequent constraint to the MYEs (and thus the implicit adoption of any errors in these two data sources)
- the assumption that the relationships between country of birth and ethnic group for the resident population at Census day hold (to an adequate degree) for international migrants
- the assumption that country of birth is an adequate proxy for nationality when assessing flows of asylum seekers

Table 1	Population estimates by ethnic group

England   Thousands

	Mid–2001	Mid–2002	Mid–2003	Absolute change 2001–03	Average annual growth rate: 2001–03 (percentages)
All people	49,450	49,647	49,856	406	0.4
White: British	42,886	42,826	42,785	−100	−0.1
White: Irish	632	622	613	−19	−1.5
White: Other White	1,348	1,398	1,438	90	3.3
Mixed: White and Black Caribbean	235	243	251	16	3.3
Mixed: White and Black African	79	84	90	12	7.0
Mixed: White and Asian	188	199	209	21	5.6
Mixed: Other Mixed	155	163	172	17	5.3
Asian or Asian British: Indian	1,052	1,079	1,113	61	2.9
Asian or Asian British: Pakistani	723	742	765	42	2.9
Asian or Asian British: Bangladeshi	283	292	302	18	3.2
Asian or Asian British: Other Asian	246	264	280	34	6.6
Black or Black British: Caribbean	576	579	584	9	0.7
Black or Black British: African	497	539	587	90	8.7
Black or Black British: Other Black	98	101	104	6	2.8
Chinese or other ethnic group : Chinese	228	259	285	57	11.9
Chinese or other ethnic group: Other Ethnic Group	224	256	278	54	11.4
Non-'White British'	6,564	6,821	7,071	507	3.8
o/w					
White: Irish and Other White	1,980	2,020	2,051	71	1.8
Mixed	657	689	723	66	4.9
Asian	2,304	2,377	2,459	155	3.3
Black	1,171	1,219	1,275	104	4.4
Chinese and Other Ethnic Group	451	515	562	111	11.6

Source: *Population Estimates by Ethnic Group, Office for National Statistics*

- that the ethnic composition of flows of migrants from each of Scotland, Wales and Northern Ireland remain the same as in 2001

More information on the assumptions underlying the estimates is available on the National Statistics website.[6]

SIZE AND CHANGE IN THE POPULATION BY ETHNIC GROUP

Table 1 sets out the estimated population by ethnic group for each mid-year from 2001–03, along with the average annual growth rate.

The mid-2001 estimates of the size of each ethnic group are reasonably close to the distribution shown in the 2001 Census results (on which the estimates are based) (see *Comparison with other sources* later in this article). The distribution changes somewhat between mid-2001 and 2003 due to differential growth.

While the large White British group showed an average annual change of –0.1 per cent over 2001–03, the figure for the non-'White British' group was 3.8 per cent. This overall growth rate covers a wide range of growth rates for the individual ethnic groups: ranging from –1.5 per cent for the White Irish to 11.9 per cent for the Chinese group. Table 2 illustrates the contributions of each component of change to the growth rates.

Whilst it is emphasised that there is uncertainty in the estimates, some patterns are evident. The White British group has small natural growth slightly outbalanced by net emigration. The White Irish group is falling due both to an excess of deaths over births and net international emigration. Growth in the Mixed groups is fuelled by natural growth. Within the Asian groups, the increase in the Asian Pakistani and Asian Bangladeshi groups is primarily caused by natural growth, whilst changes in the Asian Indian and Other Asian groups are driven by international migration. About half of the growth in the Other Asian group is attributable to flows of asylum-seekers from Iraq and Iran. Within the Black groups, Black Caribbean shows relatively low growth rates, in contrast to the Black African group, where more than a third of the estimated growth of 8.7 per cent p.a. is attributable to inflows of asylum seekers from Somalia, Zimbabwe, Democratic Republic of Congo and Sierra Leone. Very fast growth in the Chinese and Other Ethnic Group categories is almost entirely attributable to international migration, with about two thirds of the increase in Other Ethnic Group contributed by a net inflow of people born in the Philippines.

POPULATION STRUCTURE BY ETHNIC GROUP

Table 3 sets out some key indicators of population structure for each ethnic group.

The estimates show some interesting differences in sex ratios (that is, the ratio of males to females) between ethnic groups. This ratio is particularly high (showing a relatively high proportion of males) for the Other Asian group. This 'excess' is attributable to a combination of the

Table 2	Components of change, mid-2001 to mid-2003

England

Thousands/percentages

	Natural change (live births – deaths)	Migration from the rest of the UK	International migration	Average annual growth rate mid-2001 to mid-2003 (per cent)	Contributions to average annual growth rate (percentages) Natural change (live births – deaths)	Migration to/from the rest of the UK	International migration
All people	143	–37	308	0.4	0.1	–0.0	0.3
White: British	–0	–27	–66	–0.1	–0.0	–0.0	–0.1
White: Irish	–13	–0	–7	–1.5	–1.0	–0.0	–0.5
White: Other White	8	0	79	3.3	0.3	0.0	2.9
Mixed: White and Black Caribbean	17	–1	0	3.3	3.4	–0.2	0.1
Mixed: White and Black African	7	–0	5	7.0	4.0	–0.1	3.3
Mixed: White and Asian	14	–0	7	5.6	3.8	–0.1	2.0
Mixed: Other Mixed	11	–0	6	5.3	3.4	–0.0	2.0
Asian or Asian British: Indian	17	–2	47	2.9	0.8	–0.1	2.2
Asian or Asian British: Pakistani	28	–2	17	2.9	1.9	–0.1	1.1
Asian or Asian British: Bangladeshi	12	–1	8	3.2	2.1	–0.2	1.4
Asian or Asian British: Other Asian	6	–0	28	6.6	1.2	–0.1	5.5
Black or Black British: Caribbean	6	–2	5	0.7	0.5	–0.2	0.4
Black or Black British: African	19	–1	73	8.7	1.9	–0.1	7.1
Black or Black British: Other Black	5	–0	1	2.8	2.3	–0.2	0.6
Chinese or other ethnic group: Chinese	3	–0	55	11.9	0.7	–0.0	11.3
Chinese or other ethnic group: Other Ethnic Group	4	–0	50	11.4	0.8	–0.1	10.7
Non-'White British'	143	–10	374	3.8	1.1	–0.1	2.8
o/w							
White: Irish and Other White	–4	0	72	1.8	–0.1	–0.0	1.8
Mixed	48	–1	19	4.9	3.6	–0.1	1.5
Asian	63	–5	99	3.3	1.4	–0.1	2.1
Black	29	–4	79	4.3	1.2	–0.1	3.3
Chinese and Other Ethnic Group	7	–0	105	11.6	0.8	–0.0	11.0

Source: *Population Estimates by Ethnic Group, Office for National Statistics*

younger age structure of this group (and thus proportionately fewer of the predominantly female older age groups); the disproportionately 'male' component of asylum seeker inflows (two-thirds of which in 2001–03 were male) and the relatively small effect of the presence of Gurkha Armed Forces in England. Sex ratios above one are also observed for the three remaining Asian or Asian British groups; the Black African group; and the Mixed White and Asian and Mixed White and Black African groups. Again, these higher than average ratios for all these groups are partly attributable to the effect of asylum seeker inflows, and to the effect of a younger than average age structure.

Table 3 also confirms the relatively young age structure of most non-'White British' ethnic groups (though the White Irish group has both a substantially lower proportion of people aged under 16 and a higher proportion of people of retirement age).[7] It may also be noted that the White Other, Chinese and Other ethnic groups also have a smaller proportion both of people aged under 16 and people of retirement age than the White British group: a phenomenon which is consistent with recent in-migration of young adults. Not surprisingly, each of these groups also has a higher proportion than White British of women of fertile age in the population (defined here as 15–44).

The Total Period Fertility Rates (TPFRs) for 2003 shown in Table 3 are calculated in the same way as the rates for 2001 previously presented by the authors.[8] Whilst the overall rise in the TPFR between 2001 and 2003 is reflected in higher values for each group, patterns of relative estimated fertility remain similar (though not unchanged). As noted in that previous work, the differences between TPFRs for different groups are less pronounced than those estimated in other research, and this may reflect convergence of fertility rates of migrant populations to that of the indigenous population, or may be an artefact of the various estimation methodologies (further investigation of this topic is planned during 2006).

Finally, Table 3 provides an implied standardised mortality ratio for each group. It is stressed that this is not a direct estimate of the mortality of each group. The methodology underlying the estimates assumes that there is no difference between mortality rates for a given age, sex and LAD of residence for different ethnic groups. However, some ethnic groups will experience higher (or lower) mortality rates across England as a whole simply because they are concentrated in areas with relatively high (or low) mortality rates. The mortality ratios presented here, then, are simply an illustration of the effect that the geographical location of an ethnic group has on the estimated number of deaths in that group.

SUBNATIONAL PATTERNS OF DEVELOPMENT

In addition to patterns of growth seen at the England level, the estimates provide detail on developments within smaller areas, from Government Office Region (GOR) down to the level of the Local Authority District. Table 4 sets out the estimates for each ethnic group within each GOR while Table 5 provides some analysis of the results for the non-'White British' population within each GOR.

Clearly, London retains great concentrations of the non-'White British' ethnic groups, though its proportion of the total non-'White British' population has fallen from 44.7 per cent in 2001 to 42.5 per cent in 2003. Indeed, amongst the GORs, London shows the lowest annualised growth rate of the non-'White British' population over the period (although its absolute increase is more than 68 thousand). The two GORs with the highest growth rate of the non-'White British' population are those with the smallest base of that population and there is a general pattern of high growth rates for this group being associated with small base populations. The components of change columns illustrate a pattern of natural growth and strong net international in-migration of these groups in all areas, but a pattern of net internal migration from London, and

---

Table 3	Population structure by ethnic group: mid-2003

England

	Sex ratio *	Percentage under 16	Percentage retirement age	Percentage of females aged 15–44	Estimated Total Period Fertility Rate**	Implied Standardised Mortality Ratio ***
All people	96.0	20	18	41	1.73	100
White: British	95.7	19	20	39	1.73	100
White: Irish	90.4	6	32	31	1.61	101
White: Other White	91.2	14	12	57	1.60	98
Mixed: White and Black Caribbean	97.3	55	3	43	1.81	103
Mixed: White and Black African	101.1	44	3	47	2.04	105
Mixed: White and Asian	104.2	45	4	45	1.57	100
Mixed: Other Mixed	97.5	42	4	47	1.61	101
Asian or Asian British: Indian	101.1	21	9	53	1.50	103
Asian or Asian British: Pakistani	104.4	33	5	52	2.24	110
Asian or Asian British: Bangladeshi	104.3	36	5	52	2.06	112
Asian or Asian British: Other Asian	124.7	22	6	53	1.97	101
Black or Black British: Caribbean	88.4	19	14	51	1.55	106
Black or Black British: African	102.4	27	3	60	2.07	107
Black or Black British: Other Black	94.4	36	4	56	1.52	105
Chinese or other ethnic group: Chinese	96.7	15	6	62	1.39	102
Chinese or other ethnic group: Other Ethnic Group	90.2	17	4	63	1.62	100

*　Males per 100 females.

**　Total Period Fertility Rates estimated by applying differentials for England as a whole to the ASFR profile for England (including births to mothers not resident in England and Wales) for calendar year 2003 (ONS, unpublished).

***　100 x estimated deaths divided by expected deaths. Since no differentials in mortality rates between ethnic groups within an LAD/age/sex group are posited in this work, the implied SMR shown here reflects only the differences in the estimated number of deaths caused by the geographical distribution of an ethnic group (in areas of relatively high or low mortality).

Source: Population Estimates by Ethnic Group, Office for National Statistics

into other GORs. Indeed, the estimated flow of non-'White British' from London to elsewhere in England is very similar in size to the flow from outside the UK into London. Flows to other parts of the UK are relatively small, though the North East, notably, shows a net inflow of non-'White British'.

The figures shown in Table 5 illustrate the faster growth of the non-'White British' population in areas with a smaller starting population, though the East of England GOR stands out as having both a higher than average (median) proportion of non-'White British' and a higher than average growth rate for this group.

An equally revealing analysis considers change by type of local authority. For these purposes, local authorities in England are classified as London authorities; Unitary Authorities and Metropolitan Districts; or county districts.[9] Broadly, this classification distinguishes between London, urban centres outside London, and less urban areas, though it is acknowledged that this correspondence is far from being exact.[10] Table 6 provides a summary of results for each type of local authority. Comparing the top section of the table, relating to all people, and the bottom section, relating to the non-'White British' group, shows that, for each type of authority, overall population growth is outstripped by growth of the non-'White British' group. Figures for migration suggest

Table 4	Population by ethnic group: Government Office Regions, mid-2003

England Thousands

	North East	North West	Yorkshire and The Humber	East Midlands	West Midlands	East of England	London	South East	South West
Total population	2,539.4	6,804.5	5,009.3	4,252.3	5,319.9	5,462.9	7,387.9	8,080.3	4,999.3
White: British	2,422.8	6,200.6	4,540.8	3,836.7	4,522.8	4,905.5	4,383.5	7,261.2	4,711.4
White: Irish	9.5	75.7	32.9	36.1	70.4	62.3	207.1	84.1	35.0
White: Other White	26.0	89.7	66.9	67.1	76.6	157.4	614.2	245.4	94.7
Mixed: White and Black Caribbean	3.7	24.4	19.9	22.5	41.9	22.6	73.1	27.7	15.5
Mixed: White and Black African	2.4	11.2	5.2	4.5	4.8	7.9	37.3	11.9	5.0
Mixed: White and Asian	5.8	20.0	16.5	13.1	20.7	20.1	65.8	34.0	13.3
Mixed: Other Mixed	3.7	15.5	10.0	9.5	13.6	17.2	65.8	26.0	10.6
Asian or Asian British: Indian	13.1	81.7	57.6	127.8	184.4	62.8	454.2	105.9	25.1
Asian or Asian British: Pakistani	16.3	126.3	151.8	33.5	161.9	45.2	153.9	65.0	11.4
Asian or Asian British: Bangladeshi	7.0	28.9	14.1	8.5	34.0	22.6	161.0	19.0	6.4
Asian or Asian British: Other Asian	5.1	19.2	15.9	14.1	25.2	18.2	143.2	31.2	7.8
Black or Black British: Black Caribbean	2.0	23.2	22.8	29.0	83.7	33.5	337.2	36.8	15.9
Black or Black British: Black African	7.0	27.9	19.1	17.3	24.3	33.0	401.3	44.5	12.9
Black or Black British: Other Black	0.7	6.0	3.9	4.2	10.3	7.3	61.4	7.0	3.1
Chinese or Other Ethnic Group: Chinese	7.9	34.6	17.8	17.6	24.6	26.1	96.7	41.7	17.7
Chinese or Other Ethnic Group: Other	6.5	19.6	14.3	10.8	20.8	21.2	132	38.9	13.6
Percentage share of non-'White British' population in England	1.7	8.5	6.6	5.9	11.3	7.9	42.5	11.6	4.1

Source: Population Estimates by Ethnic Group, Office for National Statistics

Table 5	Non-'White British' population: Government Office Regions, mid-2003

England Thousands/percentages

Area	Total population	Non-'White British' population	Non-'White British' as percentage of population of area	Average annual growth mid-2001 to mid-2003 (per cent)		Components of change (Non-'White British') Average annual growth implied by component (per cent)			
				Total	Non-'White British'	Natural change (live births – deaths)	Internal migration	Migration to/from the rest of the UK	International migration
A North East	2,539	117	4.6	−0.0	9.9	1.2	2.9	0.4	5.7
B North West	6,805	604	8.9	0.2	5.3	1.0	1.1	0.0	3.2
D Yorkshire and The Humber	5,009	469	9.4	0.3	5.5	1.4	0.9	0.0	3.2
E East Midlands	4,252	416	9.8	0.7	5.3	0.9	2.0	0.0	2.4
F West Midlands	5,320	797	15.0	0.4	3.8	1.2	0.2	−0.1	2.6
G East of England	5,463	558	10.2	0.6	7.4	1.0	4.1	−0.0	2.4
H London	7,388	3,004	40.7	0.4	1.2	1.2	−2.7	−0.2	2.9
J South East	8,080	819	10.1	0.4	6.3	0.9	3.0	−0.1	2.5
K South West	4,999	288	5.8	0.6	9.1	0.7	5.8	0.2	2.4

Source: Population Estimates by Ethnic Group, Office for National Statistics

Table 6	Non-'White British' population: type of local authority

England
Thousands

| | Mid-2001 population estimate | Components of change 2001–2003 | | | | | Average annual growth 2003/2001 (per cent) | | | | |
		Natural change (live births – deaths)	Internal migration	Migration to/from the rest of the UK	International migration	Other	Total	Natural change (live births – deaths)	Internal migration	Migration to/from the rest of the UK	International migration
All people											
London local authority districts	7,322	98	–202	–5	182	–7	0.4	0.7	–1.4	–0.0	1.2
Unitary Authorities/Met. Districts	19,129	51	–65	–13	114	–4	0.2	0.1	–0.2	–0.0	0.3
County districts	22,998	–6	268	–19	12	3	0.6	-0.0	0.6	–0.0	0.0
Total	49,450	143	0	–37	308	–8	0.4	0.1	0.0	–0.0	0.3
Non-'White British'											
London local authority districts	2,936	69	–158	–10	171	–4	1.2	1.2	–2.7	–0.2	2.9
Unitary Authorities/Met. Districts	2,231	51	5	–3	147	–1	4.4	1.1	0.1	–0.1	3.2
County districts	1,397	23	152	3	57	4	8.2	0.8	5.3	0.1	2.0
Total	6,564	143	0	–10	374	–1	3.8	1.1	0.0	–0.1	2.8

'Other' includes changes in Armed Forces and prisoner populations, revisions made to the MYEs, and any residual effect of constraining to the MYEs.

Source: Population Estimates by Ethnic Group, Office for National Statistics

Figure 1	Population turnover: 2002–2003

Proportion of all people resident in 2002 or 2003

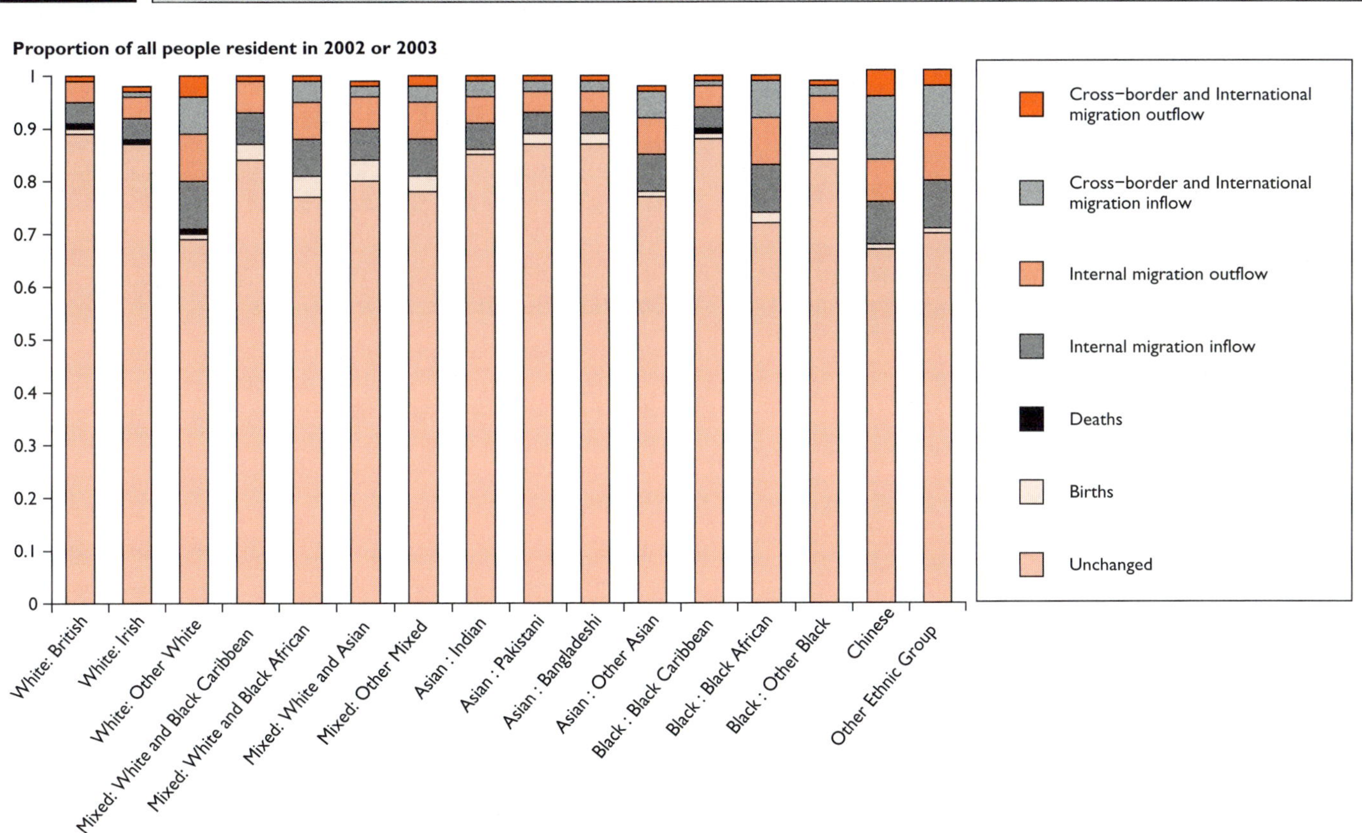

that there is generally a pattern of non-'White British' population growth in London and in the Unitary Authorities and Metropolitan Districts being driven by international in-migration, while growth in the county districts is largely attributable to migration from those, typically more urban, areas.

Analysis can also be carried out at the level of the individual LAD. Identification of the LADs with the fastest growing non-'White British' population is of limited interest – high growth in East Cambridgeshire, for example, is largely attributable to changes in US Armed Forces (primarily assigned to the White Other category) rather than a more general demographic phenomenon. In general, the highest growth rates are seen in those areas with small starting populations of non-'White British', with, conversely, the lowest growth rates associated with high proportions. Thirteen LADs, all in Inner London, and Inner London itself, show a fall in the proportion of the total population belonging to a non-'White British' group. This reflects a rise in the number of White British people, due largely to natural growth, but also partly due to net migration from elsewhere in England and the rest of the UK (in contrast, all other ethnic groups showed net flows from Inner London to elsewhere in England and the rest of the UK). These changes in Inner London may be contrasted with the results for London as a whole discussed previously.

POPULATION TURNOVER

Population turnover refers to the degree to which a population gains or loses individuals over time. It is distinct from growth as it is not the aggregate increase or decrease that is of interest, but rather the number of individuals who join or leave the population (so, for example, an equal increase in migration both into and out of an area would leave the overall population unchanged but would increase the degree of population turnover). One of the advantages of adopting a cohort-component methodology, rather than relying on estimates based on sample surveys, is that examinations of population turnover flow naturally from the results.

Figure 1 illustrates patterns of population turnover by ethnic group for 'an average LAD in 2002 to 2003'. It considers all people who were resident in such an LAD in 2002 or 2003 and shows whether they were present in both years (that is, unchanged); or whether, and how, they entered or left the population over the period. Two points may be made in relation to the chart.

Firstly, high growth rates are not synonymous with high turnover. For example, the second highest turnover (after Chinese) is seen in the White: Other group, which had the fourth lowest growth rate over 2002–03. High turnover in this group is a result of relatively high IPS migration flows and a substantially higher propensity to migrate within England. It is worth remarking that the high IPS migration flows for this group are substantially due to flows to and from Australia, New Zealand, Canada, South Africa and EU countries, and that the much discussed flows from Eastern Europe have a relatively small impact on the estimates for 2003 (though it seems likely that the net flow from Eastern Europe will play a more significant role in estimates for 2004 and subsequent years following the expansion of the EU).

Secondly, the extent of turnover is, for all groups other than Chinese, more dependent on internal migration than cross-border or international migration, and that the propensity to migrate within England varies substantially between ethnic groups, with high rates seen for the Black African and White Other groups but low rates for the Asian Pakistani, White British and Asian Bangladeshi groups. The low rate for Asian Bangladeshi is of particular interest as the young age structure of this group would suggest that crude propensities to migrate would be higher than average.

DIVERSITY AND SEGREGATION

Measurements of 'ethnic diversity' are currently attracting much attention. The often-used measure of 'the proportion of an area's population not from the White British group' is obviously flawed – an area composed entirely of White Irish, for example would be shown as being of maximum ethnic diversity while in reality there is no ethnic diversity at all (with respect to the Census ethnic groups). An intuitively appealing measure is the probability that two people drawn at random from the population of the area are of different ethnic group (In practice, the two measures discussed will produce a similar ranking of LADs in England except for some changes in order for the more ethnically diverse areas in London).

Figure 2 shows the distribution of the ethnic diversity measures for LADs in England in 2003. Not surprisingly, the most ethnically diverse LADs are concentrated in London, with Birmingham and Leicester also showing indices above 0.50. Other higher than average indices are seen around the former Metropolitan District of Greater Manchester, and an interesting pocket of higher than average indices is found in the East of England, where Forest Heath and East Cambridgeshire, both of which contain significant numbers of US Armed Forces, abut Cambridge, with its presence of overseas students.[11]

A different approach to considering the spatial distribution of ethnic groups is to consider segregation. One measure of segregation is the Index of Dissimilarity described in Box 2.

Table 7 sets out the dissimilarity indices for each ethnic group for LADs in England[13]. The indices suggest that the Asian Pakistani, Asian Bangladeshi, Black Caribbean and Black African groups all have similar degrees of segregation, with the Mixed groups and Chinese having the lowest. Care should be taken in interpreting these indices however – as they are calculated at the level of the LAD they will not reflect the degree of segregation within an LAD – the geographical level for which, it can be argued, such measures are of most interest.

COMPARISON WITH OTHER SOURCES

Figure 3 provides a comparison of population estimates by ethnic group for 2001 taken from the 2001 Census, the Labour Force Survey (LFS), and the Population Estimates by Ethnic Group (as the LFS does not separately identify White: Irish, this group has been added to 'White: Other' for the Census and Population Estimates by Ethnic Group shown in the figure).

Unsurprisingly the Census and the (census-based) Population Estimates show similar distributions, though the adjustments for under-enumeration made in the Mid-Year Estimates (which are reflected in the Population Estimates by Ethnic Group) do have a proportionately greater effect on those ethnic groups with relatively more young men and a greater concentration in urban areas.

Some reasons for differences between census-based estimates and LFS results have been discussed by Heap[14]. In addition to the sampling variability inherent in LFS estimates for small groups, the differences in the estimates for the Mixed groups between the census-based estimates and the LFS may also be partly attributable to differences in coding rules for the two data sources. The need to aggregate the White Irish group for the purposes of comparison, as mentioned above, may explain the discrepancy in the White Other estimate.

Differences between the LFS and the census-based estimates are more striking when change over time, rather than the absolute level of population, is considered. Figure 4 illustrates estimated change between 2001 and 2003 using the Population Estimates by Ethnic Group and the

<table>
<tr><td>Figure 2</td><td>Ethnic group diversity indices, local authority districts, 2003</td></tr>
</table>

England

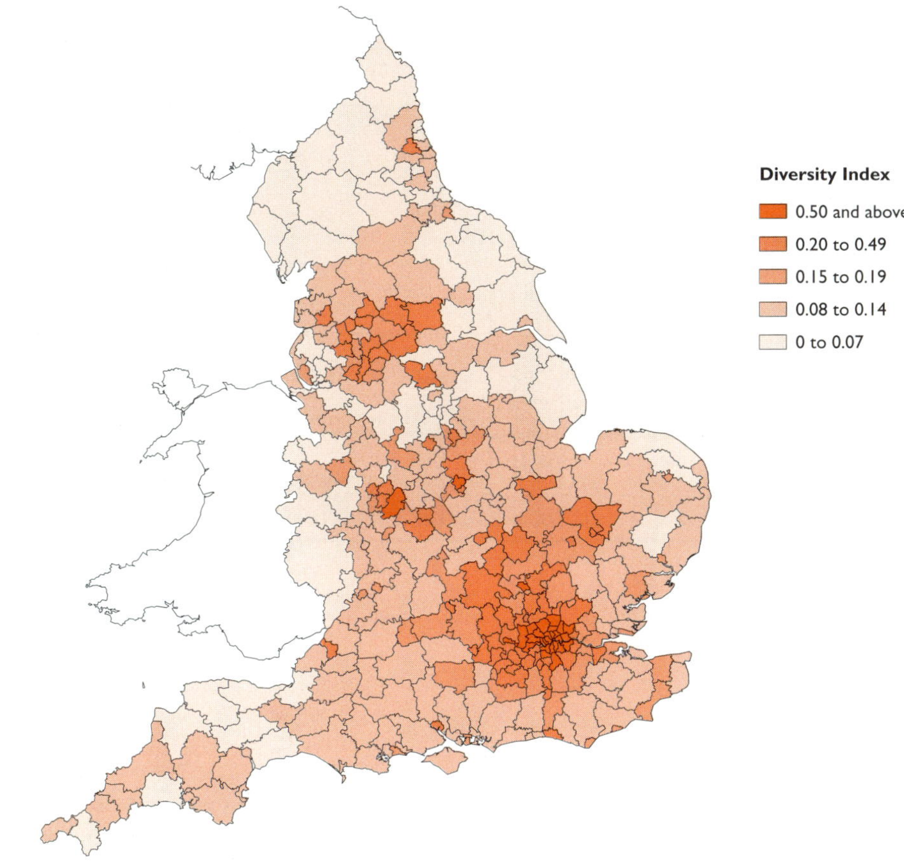

Diversity index: Probability of two persons selected at random in a local authority district belonging to different ethnic groups

Source: Population Estimates by Ethnic Group, Office for National Statistics

Box two

Dissimilarity Index

The Dissimilarity Index[12] D can be intuitively understood as the extent to which a population group is not distributed out evenly amongst the population as a whole. It is defined, for this geography, as:

$$D = 1/2 * \Sigma \mid (x_i/X - y_i/Y) \mid \text{ where}$$

x_i = number of that ethnic group in LAD i;
X = total number of that ethnic group in England;
y_i = number of people not in that ethnic group in LAD i
Y = total number of people not in that ethnic group in England.

The Index can take values between zero and one. A Dissimilarity Index of zero indicates that a group is distributed evenly amongst the population as a whole. An Index of one indicates that a group is located only in areas which contain no members of other groups.

<table>
<tr><td>Table 7</td><td>Dissimilarity indices for ethnic groups, mid-2003</td></tr>
</table>

England

Ethnic group	Dissimilarity index
White: British	0.44
White: Irish	0.29
White: Other White	0.35
Mixed: White and Black Caribbean	0.32
Mixed: White and Black African	0.33
Mixed: White and Asian	0.24
Mixed: Other Mixed	0.29
Asian or Asian British: Indian	0.51
Asian or Asian British: Pakistani	0.56
Asian or Asian British: Bangladeshi	0.56
Asian or Asian British: Other Asian	0.44
Black or Black British: Black Caribbean	0.56
Black or Black British: Black African	0.57
Black or Black British: Other Black	0.53
Chinese or Other Ethnic Group: Chinese	0.31
Chinese or Other Ethnic Group: Other	0.39

Source: Population Estimates by Ethnic Group, Office for National Statistics

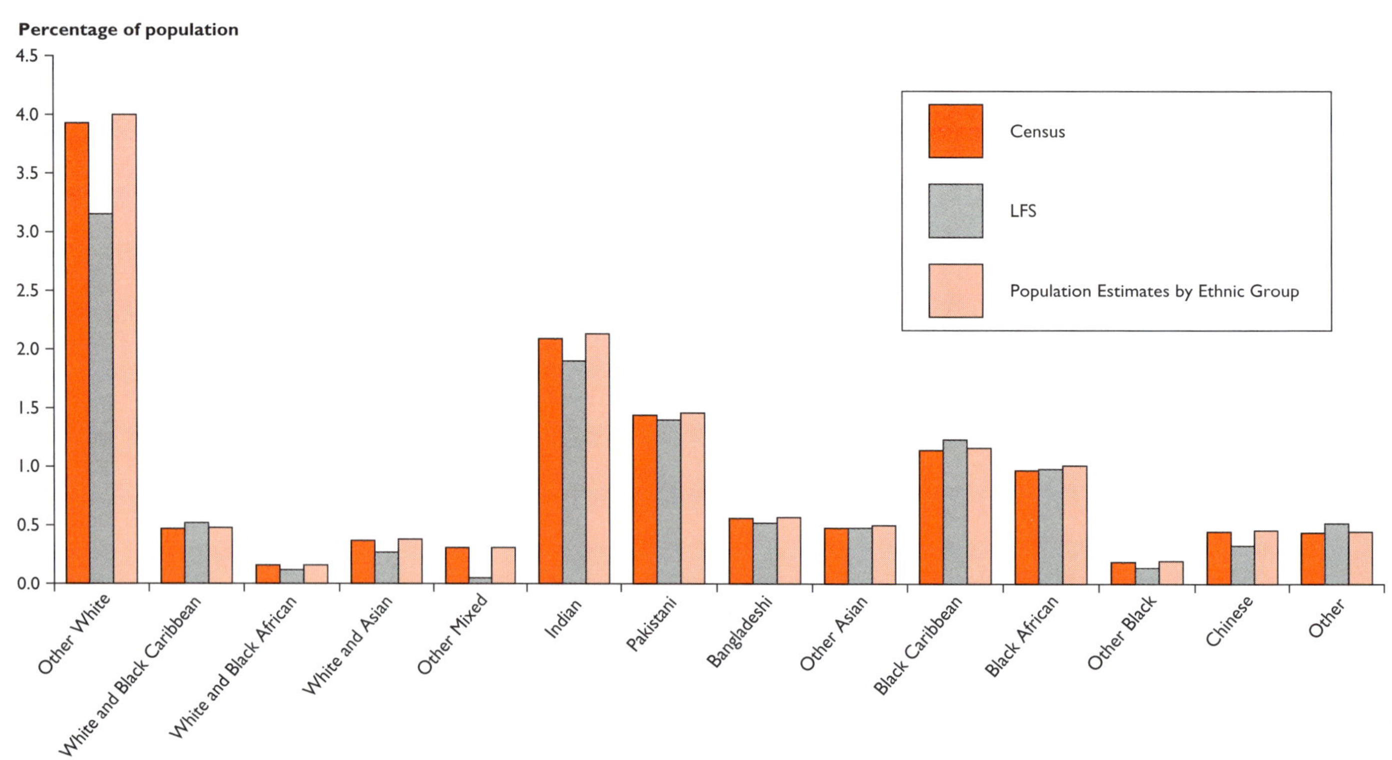

Figure 3 | Population by ethnic group: England, 2001

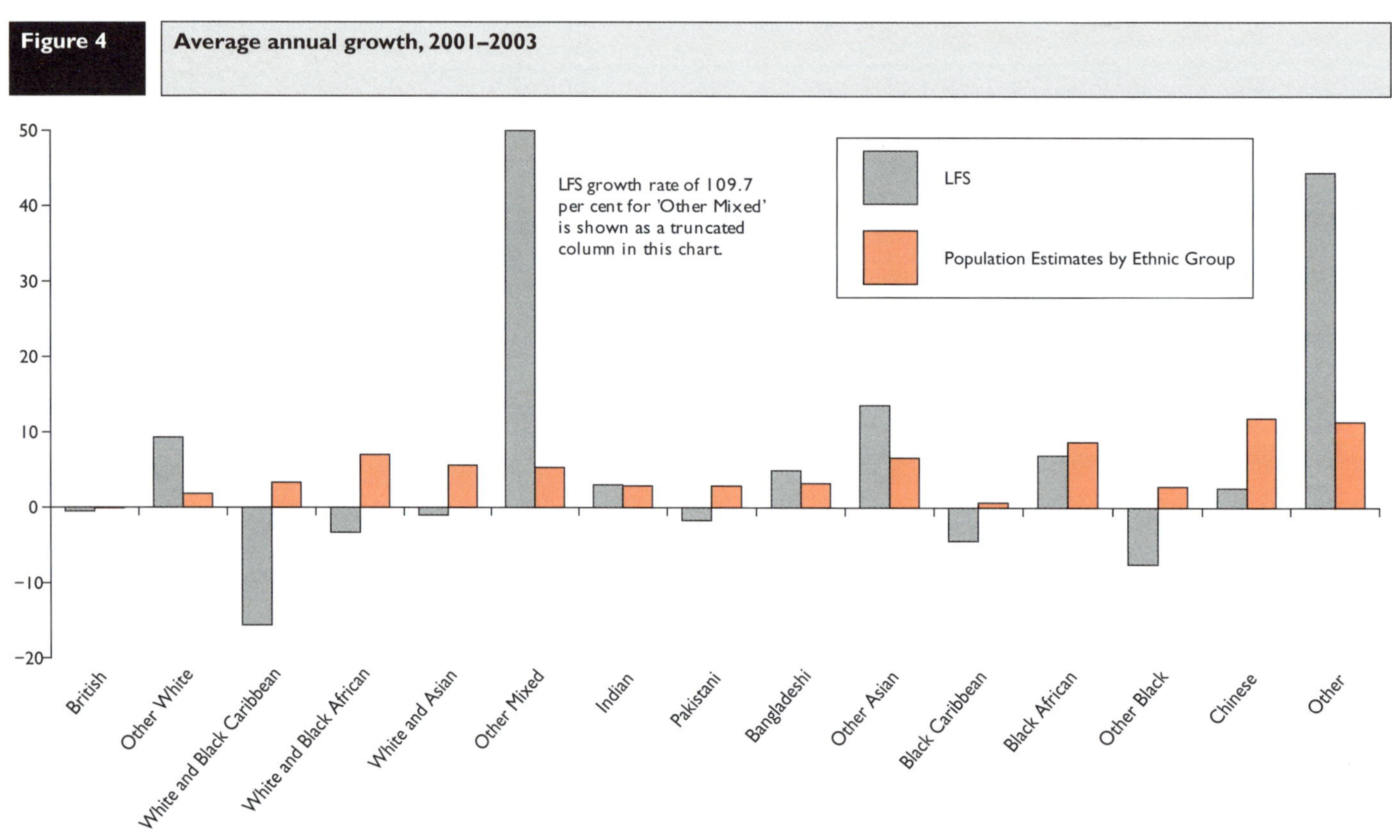

Figure 4 | Average annual growth, 2001–2003

LFS. Not surprisingly, the cohort-component method of the Population Estimates produces growth rates which accord more with expectation than the comparison of the two LFS estimates (both of which are subject to sampling error) which show, for example, a decrease in the Asian: Pakistani population over the period.

The Annual Labour Force Survey also provides estimates of the population by broad (that is, 5 way) ethnic group for LADs and higher administrative areas. Inevitably, however, the geographical concentration of the non-'White British' groups means that reliable estimates are not possible for the majority of combinations of groups/areas.[15]

MORE INFORMATION

The full set of published estimates, together with further explanatory material, is available on the National Statistics website at www.statistics.gov.uk/StatBase/Product.asp?vlnk=14238.

CONCLUSION

The population estimates by ethnic group for mid-2001 to mid-2003 provide more up-to-date figures than the 2001 Census and are more reliable than sample surveys such as the LFS when considering small groups or areas, or change over time. However, it is stressed that the estimates are experimental, and that the assumptions underlying the estimates should be taken into account when interpreting the figures.

Whilst ease of exposition within this article has frequently led to the presentation of statistics for the non-'White British' group (actually composed of 15 groups from the National Statistics classification of ethnicity), it is stressed that the estimates, and the research on which they are based, show that it is quite misleading to assume homogeneity of demographic characteristics within this 'group'.

Key findings

- The non-'White British' population grew at an estimated annual rate of 3.8 per cent between 2001 and 2003, compared to a rate of −0.1. per cent for the White British group.
- The fastest growing groups were Chinese, Other ethnic group, and Black African.
- Growth of non-'White British' outside London is driven by migration within England; within London, by international migration.
- Population 'turnover' within LADs is highest for the Chinese and White Other groups, and lowest for the White British and Black Caribbean groups.
- Asian Pakistani, Asian Bangladeshi, Black Caribbean and Black African groups are relatively highly segregated by LAD with the Mixed and Chinese groups having lower segregation by LAD (this measure does not provide information about segregation within LADs).

NOTES AND REFERENCES

1. Scott A, Pearce D and Goldblatt P (2001) The sizes and characteristics of the minority ethnic populations of Great Britain – latest estimates. *Population Trends* **105**, pp 6–15.
2. Schuman J (1999) The ethnic minority populations of Great Britain – latest estimates. *Population Trends* **96**, pp 33–43.
3. Large P and Ghosh K (2006) A Methodology for Estimating the Population by Ethnic Group for areas within England. *Population Trends* **123**, pp 21–31.
4. Office for National Statistics (2003) *Ethnic group statistics: A guide for the collection and classification of ethnicity data*. The Stationery Office: London.
5. Office for National Statistics (2004) *National Statistics Code of Practice: Protocol on Data Presentation, Dissemination and Pricing*. The Stationery Office: London, available at www.statistics.gov. uk/about/national_statistics/cop/protocols_published.asp.
6. Office for National Statistics. *Population Estimates by Ethnic Group: Issues and Guidance*, available at www.statistics.gov.uk/StatBase/ Product.asp?vlnk=14238
7. A relevant discussion of the age structure of in-migrant populations (that is, people born outside the UK) is provided by Rendall M and Salt J *Focus on People and Migration* chapter 8, available at www. statistics.gov.uk/statbase/Product.asp?vlnk=12899
8. Large P and Ghosh K *Ibid*. p 24.
9. See also Stillwell J and Duke-Williams O *Ethnic population distribution, immigration and internal migration in Britain: what evidence of linkage at the district scale?*, paper prepared for the British Society for Population Studies Annual Conference, September 2005. The further distinction between Unitary Authorities and Metropolitan Districts used by Stillwell and Duke-Williams is less important when looking at England, as here, rather than Great Britain as a whole.
10. Following the definitions used in the Rural and Urban Classification 2004, the 2001 Census showed that 81 per cent of people in England lived in an urban area in 2001. Though 65 per cent of people living in a county district lived in an urban area, the corresponding proportion for Unitary Authorities/Metropolitan districts was 92 per cent, and that for London LADs 100 per cent. Source: Table KS01, *Key Statistics for rural and urban classification 2004* available at www. statistics.gov.uk/census2001/product_ks_ur_2004.asp.
11. Office for National Statistics (2006) The UK population at the start of the 21st century. *Population Trends* **122**, p 13.
12. The interpretation of the Dissimilarity Index, and other indices relating to segregation, is discussed in, for example, Simpson L (2004) Statistics of racial segregation: measures, evidence and policy, *Urban Studies* **41**, pp 661–681; and Johnston R, Poulsen M and Forrest J (2005) On the measurement and meaning of residential segregation: A response to Simpson, *Urban Studies* **42**, pp 1221–27.
13. Similar indices for types of LAD in the UK based on 2001 Census data are provided in Stillwell and Duke-Williams, *Ibid*.
14. Heap D. *Comparison of 2001 Census and Labour Force Survey labour market data*, available at www.statistics.gov.uk/cci/article. asp?ID=1037
15. See Supplementary Table 1.1, Annual LFS 2002/03 available at www.statistics.gov.uk/statbase/Product.asp?vlnk=13210

Population bases and statistical provision: towards a more flexible future?

Chris W Smith and Julie Jefferies
Population and Demography Division,
Office for National Statistics

In an increasingly complex and mobile society, there is a need for population estimates to be produced on a more flexible basis. Different uses of population data may require information to be output on different population bases, such as where people usually live ('usual residence') or where they are on a particular day ('population present'). This article explores many of the issues associated with defining 'the population'.

Following consultation with data users, it outlines recommended population definitions that could facilitate a more flexible approach. Comparisons are made between the output bases produced from the 2001 Census and the more adaptable outputs potentially available in future if the recommended definitions were used.

BACKGROUND

In 2003 the Office for National Statistics (ONS) published *A Demographic Statistics Service for the 21st Century*.[1] The report was intended to stimulate discussion about the strategy for population measurement over the next few decades. It addressed the issue of providing population statistics based on different definitions (see Box One) to reflect diverse and dynamic living patterns in modern society. Increases in multiple residency, weekly commuting to work, numbers of children of parents living apart and more frequent movements between countries all reflect these more complex lifestyles. The report stressed the need to be more flexible in defining residency, while encouraging ONS to explore various definitions of the population from existing sources. No single population definition can meet all user needs: although the usually resident population is important, some service providers also require counts on other bases.

In addition to usual residence, the report listed 14 other output bases of potential interest (Table 1). Only three of these were produced and published from the 2001 Census, with two more theoretically or partly available. The three were usual residence, household population and communal population, with out-of-term population and work-day population being theoretically or partially available.

The key to taking this innovative approach lay in asking major data users which output bases they would find most useful without predefining their options. They were also asked how ONS should strike a balance between improving the quality of outputs on the usual residence base and producing new types of population output and analysis. This would include discussion of which working definitions would deliver their requirements in providing effective measures of living patterns in modern

Box one

What are population definitions and bases?

At any one time, the 'population' of an area could include people who live there normally, those living there for part of the time, those living there temporarily, those visiting the area, those working in the area and those who normally live in the area but are currently somewhere else. The numbers in each of these groups are likely to vary depending on the time of day, day of the week and time of year. Therefore it is important to define exactly what we mean by 'the population' of an area when collecting information from or producing information relating to that population.

When deciding how to define a country's population, the key issues are who to include in the population and where to include them:

Who should be included as part of the country's 'population'?

One approach is to include all people living or staying in the country on a particular day or night. This is commonly known as defining the population on a 'population present' or *de facto* basis. However, this includes visitors (those who are simply in the country for a short time) as part of the country's 'population' and excludes those staying abroad for a similar period. This issue can be avoided by using a 'usual residence' base, that is to say including all people who 'usually' live in the country. This requires a clear definition of 'usual residence', a concept discussed in detail in this article. For some purposes it may be important to include visitors as well as usual residents in a population base as they may use services or contribute to the economy.

A key issue in distinguishing usual residents from visitors is the amount of time spent in the country.

Where in the country should each person be included?

Using a 'population present' base, people are included wherever they are staying on a particular night, whereas a 'usual residence' base includes people in the area where they usually live. This article discusses some of the difficulties in determining where certain groups of people with more than one address or no address 'usually' live. For some areas, the population on a 'population present' base can be very different from that on a 'usual residence' base.

The population base is simply the way in which the population is defined for the purposes of collecting or producing data. Table 1 shows a variety of possible bases. The population base at enumeration (when the data are collected) may not necessarily be the same as the population base for outputs (when the data are published).

Table 1 — Possible output bases for population and actual availability from the 2001 Census

Output Base[1]	Available from 2001 Census?
1. **Usual residence** – the population at the address at which they *usually* live	Yes
2. **Household population** – the population living in private households	Yes
3. **Communal population** – the population living in communal establishments	Yes
4. **Out of term population** – the population usually resident in an area out of term-time	Theoretically available but in practice not produced
5. **Seasonal population** – either the usually resident population at different times of year or enhanced to include visitors (for example, tourists)	No
6. ***De facto* population** – population present on Census night	No
7. **Legal population** – any 'legal' definition of the population to serve a particular purpose	No
8. **Week-day population** – population usually resident/present overnight during the week (for example, Monday to Thursday nights)	No
9. **Weekend population** – population usually resident/present at the weekend (for example, Friday night to Sunday)	No
10. **Temporarily resident population** – those present in an area but are not usually resident there (for example, divided children, temporary migrant workers)	No
11. **Work-day population** – population present in an area during normal working hours Monday to Friday	Workplace population and daytime population estimates (workplace population plus those resident in the area who are not working) published for ages 16–74. Place of study not available
12. **Average population** – average of various other population bases	No
13. **UK residents living abroad** – temporarily or for part of the time (for example, international commuters)	No
14. **Non-UK residents living in the UK** – for example, visitors, short-term migrants, international commuters	No
15. **Bespoke service populations** – target population for a specific service/policy	Determined by nature of output needed

1 As described in *A Demographic Statistics Service for the 21st Century*.[1]

possible to minimise respondent burden). ONS will need to review these requirements, assess the feasibility of collecting the desired information and prioritise each requirement through further discussion with users.

USER CONSULTATION AND ENUMERATION BASES

ONS conducted a written consultation exercise on population bases[2] with users of population data over the period June to September 2004. This revealed support for output data to be based primarily on usual residence for both individuals in households and in communal establishments. Only by employing a precise and detailed definition of *usual residence* might hard-to-define groups be 'captured'. Moreover, users recommended that visitors should be recorded for two reasons. Firstly, visitors from overseas working within the UK often contribute to the economic activity of the country even though they may not be 'usually resident'. Secondly, if visitors are required to complete a census questionnaire, people may be less likely to avoid their legal obligation to comply with the census on the grounds that they do not have a 'usual' residence.

As a follow up to the consultation, ONS convened the Population Definitions Working Group in September 2004 with a brief to provide definitional advice on population data issues. The working group

society. This article addresses these issues and analyses data users' views in order to provide guidance for a more flexible and effective future population provision. This discussion has very broad statistical relevance but in the short-term will be focused on the 2011 Census which is likely to underpin population estimates during the following decade.

It is important to note that the issues discussed in this article reflect users' requirements, but in practice, these will be constrained by operational considerations on the 2011 Census (for example, space on the questionnaire and the need to keep the enumeration process as simple as

consisted of population data experts from a variety of backgrounds (local authorities, central government, academia and the private sector) and a core of ONS topic experts. Its role included addressing specific issues in preparation for the 2011 Census. The role was purely advisory, since final responsibility for the census lies with ONS.

The working group endorsed the conclusions of the 2004 consultation. While recognising that the choice of output base was a key issue, it noted that this would impinge on the choice and definition of enumeration base. Accordingly, the group supported the use of a *usual residents plus visitors* enumeration base in the 2011 Census as meeting key user needs. ONS has since agreed that the 2011 Census should have this composite *usual residents plus visitors* enumeration base. Discussions of output bases for census data for 2011 were set against this agreed enumeration base, though ONS has yet to finalise precise definitions of both *usual residence* and *visitors* for the 2011 Census. The sections below summarise the main definitional issues discussed by the working group.

USUAL RESIDENCE

Usual residence is where people usually live. For most people this is straightforward because they normally live at one address only. However, difficulties in defining usual residence arise in some cases, particularly with the more mobile sections of the population. For example, a retired couple sharing time between their UK home and another home in Spain: are they *usually resident* in the UK? Similarly, a person with a home in Berkshire who spends the weekend there with their family, but also has a weekday flat in London near to their workplace: deciding where this person is *usually resident* is fraught with difficulty. They may spend more time in the London flat, but their family lives at their Berkshire home: which of the two is their *usual residence*? Two key issues require resolution: first, distinguishing UK residents from those usually living abroad, and, second, assigning UK residents to local areas within the UK.

In 2001 the Census definition[3] of a usual resident was:

'... someone who spends the majority of their time residing at that address'.

There were a number of clarifications to this definition (see Box Two). These were stated clearly on the 2001 Census questionnaire.

Box two:

2001 Census definition of a usual resident

A usual resident is generally defined as someone who spends the majority of their time residing at that address. It includes:

- people who usually live at that address but who are temporarily away from home (on holiday, visiting friends or relatives, or temporarily in a hospital or similar establishment) on Census day
- people who work away from home for part of the time, or who are members of the armed forces
- a baby born before 30 April 2001, even if still in hospital; and
- people present on Census day, even if temporarily, who have no other usual address

---

The enumeration base for the 2001 Census was usual residents only, allowing some individuals to avoid the Census by claiming that they had no 'usual' residence. The composite base of usual residents plus visitors agreed for 2011 could help to overcome this problem.

Although some people have more than one residence, for the purposes of the census each person must be assigned only one *usual* residence. Those people without any usual residence are assigned to an address to ensure that they are not omitted from the count; this would be where they are on Census night.

The working group recommended a series of clarifications in defining usual residence. Persons working away from home during the week and returning to the permanent or family home at the weekend should have the permanent or family home recorded as their usual residence, even if the majority of their time is spent at their 'working week' address. This enables family relationship data to be obtained. It should be noted that other users might prefer such people to be counted where they live during the working week; this is discussed later in the section on 'secondary residence'.

Armed forces need to be treated differently from weekly commuters, as they frequently work away from their permanent or family residence for long periods of time. In the 2001 Census members of the armed forces with partners were counted at their family residence but single service people were counted at their base address. The usual residence of armed forces is a contentious issue, since different users require different definitions. To maintain consistency with population estimates' methodology, some users prefer the usual residence to be the address at which forces' personnel live when working at their base, as this is where more resources are used (this would be a change from the 2001 definition). Other users prefer armed forces' personnel to be counted at their family residence so that household structures are maintained and family relationships accurately represented.

The issues discussed above also apply to people sharing their time between the UK and abroad. In general people will be considered usually resident in the UK if they spend the majority of their time in the country or if their family or permanent home is in the UK although they work abroad for part of the time. In practice 2011 Census respondents will define themselves as either usual residents or as visitors.

Consideration was given as to whether a standard duration should be applied for residency in a country, such as the 12 months recommended by the UN Economic Commission for Europe.[4] In particular, the working group considered a 'six month rule' to determine usual residence, as in the formal 2001 definition. For example if a person had spent more than six months of the past year in the UK, their usual residence would be the UK; conversely if they had spent more than six months of the past year outside the UK, they would not be usual residents (but would need to be enumerated as visitors). However there are practical difficulties with this. For example a person who moved permanently to the UK one month before Census day would be captured as a visitor rather than a resident under this definition.

This illustrates how usual residence status may be based on intent as well as past behaviour. A person who left the UK to live in another country one month before the census for short-term employment purposes might consider themselves to be usually resident in the UK if they intended to return within a couple of months, but not if they planned to stay abroad for some time. Therefore a suitable definition of usual residence in the UK rather than abroad would need to be very detailed and thus less likely to be read by the majority of respondents. Moreover, if people who are abroad on Census day have immediate family in the UK they are more likely to be included on a questionnaire by other household members than those without close family in the UK.

Another group with more than one residence in the UK is those children of parents living apart who spend part of their time staying with each parent. In such cases, the usual residence would be the address at which children spend the majority of their time. The 'tie breaker' for children divided equally between parents could be where the child is on Census night.

For students, the group recommended that usual address is their term-time address, since most students will spend the bulk of their time there. This is consistent with the 2001 Census definition, when students were recorded at their term-time address for the first time, and also consistent with population estimates.

The usual residence of those living in communal establishments needs to be clearly defined to avoid counting people both at the communal and at any family or permanent address, or not counting them at all. Users recommend that six months is an appropriate length of time to use to define usual residence in a communal. If the person has already spent or intends to spend six months or more in the establishment then usual residence would be there. Otherwise the person would be enumerated as a visitor in that establishment and as usually resident at their family or permanent address. Intended stay is necessary in defining usual residence for this group since some moves into communals may be permanent, for example an older person moving to a residential home. In such a case it would not make sense for them to be recorded as usually resident at their family or permanent address purely because they had lived in the establishment for less than six months.

Prisoners serving sentences in prison can be treated as residents or visitors in the same way as other people living in communal establishments for census purposes. However prisoners on remand are an exception to this rule since their stay in the prison is considered temporary. The group recommended that remand prisoners be recorded as usually resident at their family or permanent address, unless they have no other usual residence. Defining the usual residence of sentenced and remand prisoners separately in this way is consistent with the 2001 Census and with mid-year population estimates.

Those usually resident in the UK but with no usual address were enumerated where they were present on Census night in 2001. The group recommended a similar practice in 2011. These 'resident visitors' are of particular interest as they are likely to be mobile and hard to count. Users want resident visitors to be separately identifiable in outputs to gain a greater understanding of the numbers and characteristics of these people. For example, it would be useful to differentiate between resident visitors counted in the communal population (including those sleeping rough and in hostels), and those living in a household on Census day.

In looking at these recommendations, ONS must ensure that these hard-to-define groups are captured without unduly compromising on the majority's understanding of the definition of usual residence.

Visitors

The 1991 Census collected information on both usual residents and visitors present on Census night. In 2001 space was included on the census questionnaire to list any visitors at the address, but this visitor information was not collated. In 2011 ONS plans to use a combined enumeration base of usual residents plus visitors to provide more comprehensive coverage of the whole population.

A variety of user requirements for visitor data were identified by the public consultation.[5] Collection of visitor data would enable estimates to be made on a 'usual residents plus visitors' base which would reflect more closely the number of people using services in each area, including

visitors. This information would permit more appropriate levels of service provision and resource allocation. Although some users only require data on particular types of visitor, the most appropriate strategy is to record all visitors present on Census night, because enumerating some visitors and not others would be too complex. The collection of visitor data might also offer some operational benefit if it could be used to help estimate any undercount.

It is not feasible to collect a large amount of data from visitors since emphasis on visitors would reduce the quantity of data that could be collected from usual residents. There is also an issue of respondent burden as this group may be particularly unwilling to answer a long set of questions. Local authorities would like information on the age and sex of visitors plus an indication of usual residence (postcode if usually resident in the UK or country if usually resident abroad).

The collection of usual address information from visitors would enable two types of visitor to be distinguished:

- those with a usual address elsewhere in the UK
- those with a usual address abroad

Those with no usual address may consider themselves to be visitors at their Census day address, but should be enumerated as usually resident at that address and would be referred to as 'resident visitors'.

Visitors with a usual address elsewhere in the UK might include holidaymakers, those staying with friends or relatives for the night and people working away from home temporarily. These people should be recorded as visitors at their Census night address but also be required to complete a questionnaire or have it completed on their behalf at their usual residence. All data users want outputs on the usual residence base to include such temporarily absent persons counted at their usual address. Collecting basic demographic data from this group at their address on Census night would provide limited information on UK overnight visitors to different areas, which could further inform resource allocation. It would also enable visitors who were not counted at their usual address to be matched back to their usual address if resources permitted.

It is unlikely to be feasible to produce estimates on a 'population present' base from visitor information, since all visitors would need to be matched successfully to their usual residence and removed from that address (if in the UK) – this would be highly dependent on the quality of visitor information collected and on processing resources available.

Visitors with a usual address abroad might include holidaymakers, temporary workers from abroad, people who have retired abroad and students on short courses. Counts of all such visitors are required to inform resource allocation and service provision. In addition, users of labour market statistics have requested counts of temporary visitors from abroad who are working in the UK. Adding such people to the usually resident population base could provide a more appropriate denominator for employment rates and similar economic measures. This would require the collection of data on economic activity from visitors.

Secondary residence

Some people divide their time between several addresses. The issue of which of these addresses should be considered as the usual residence has been discussed earlier. Collecting information about secondary residences in addition to usual residence would create the potential to produce estimates on a wider range of output bases, as well as adding to understanding of the movements of particular population groups. Collection of data for multiple secondary residences would be desirable, though in practice cost and respondent burden would limit questions to one secondary residence per person.

In some cases a person's secondary residence is simply the conventional 'second home' – a weekend or holiday property owned by that person. But the issue of secondary residence encompasses much more than this. The concept of secondary residence applies to a person rather than a dwelling. For example a person's secondary residence might be the B&B where they stay from Monday to Friday while working away from home. It might be a student's vacation address or the other parent's residence, for children whose parents live apart. A particular dwelling may be the usual residence of one person and the secondary residence of another. However, a building owned by one person, but rented out to others, is not considered to be the owner's secondary residence.

Users require data on conventional second homes to understand housing demand and affordability and to inform housing provision in local areas. In addition users would like information on all types of secondary residence for service provision and resource allocation purposes. Some types of local area, in particular those with large numbers of workers staying during the week who are usually resident elsewhere, would welcome this information about their weekday populations. Collecting information on both term-time and vacation addresses for students (as in 2001) and on secondary residence for armed forces' personnel could enable both sending and receiving authorities to have a greater understanding of fluctuations in their populations.

Secondary residence information could also be used to obtain a greater understanding of the fluidity of the population, not only in specific areas but for specific groups of the population. For example, the data could provide a valuable insight into the prevalence of weekly commuting. Similarly, no estimates are available at present on the numbers of children moving regularly between the homes of parents living apart.

An additional benefit from collecting secondary residence data would be to improve the accuracy of journey to work data. Consider a person who is usually resident in Hampshire at weekends but lives and works in London during the week. Users would require information on the daily journey to work from the home in London to the workplace. However if only usual residence was collected, the point of origin for their daily journey to work would be incorrectly recorded as Hampshire. Overall this would give a misleading view of transport patterns. Thus secondary residence data may improve the accuracy of journey to work data by increasing the likelihood of identifying the correct origin of the daily journey.

In addition to these uses, secondary residence data has the potential to be used to help estimate over-count in the census by identifying people incorrectly counted as usually resident at two addresses. For example the Longitudinal Study has provided evidence that in 2001 some children were double counted at both parents' addresses.[6] A suitable method to estimate over-count would need to be developed for this potential to be realised.

Users emphasised the importance of secondary residence information in providing the flexibility to produce outputs under different residence definitions. Collecting these data would provide the ability to output populations on separate weekday and weekend bases, as well as the out-of-term population base, providing the data were of sufficient quality. It would also be a step towards the estimation of the 'average population' of local areas. The quality of these output bases would depend on how much information it is feasible to collect on the purpose and length of time spent at secondary residences and on the quality of responses to these questions. There is also the related issue of competition for space on the census questionnaire.

To obtain the information required by users, data would need to be collected not only on the second address itself but on the type of secondary residence, for example whether it is used for holidays, during the working week or is the other parent's address. In addition, those with secondary residences would ideally need to be asked how frequently they move between addresses or how much time they normally spend at each address. For those whose secondary residence is outside the UK, only the country of residence is required.

HOUSEHOLDS AND COMMUNAL ESTABLISHMENTS

Most people in England and Wales live in private households rather than in large establishments such as student halls of residence, nursing homes and army barracks. In 2001, 98.2 per cent of people in England and Wales were living in private households, while 1.8 per cent (over 934,000 people) were resident in communal establishments.[7] The latter included people living in medical and care establishments, prisons, defence and educational establishments.

The definition of a communal establishment distinguishes the communal establishment population (those living in communal establishments) from the household population (everybody else). Therefore in order to identify the household population, it is essential to define clearly a communal establishment.

In previous censuses, selected output tables were produced separately for the household population and for the communal population. There is a continuing need for these two population bases and also for information to be collected from individuals as part of household units or communal establishments. For example, national and local household projections used in planning of housing provision require a household population base and information on household units derived from the census. The household population is also used for weighting various household surveys which do not cover communal residents. In addition to these key requirements, many users need data on the characteristics of households (for example family relationships within households, household income) or of people living in different types of communal establishment (for example, age, gender, general health).

Communal establishments

In 2001, a communal was defined as:

'an establishment providing managed residential accommodation.'

In this context, 'managed' means that the accommodation is supervised on a full-time or part-time basis.

Users emphasised the importance of consistency in the definition of a communal establishment and agreed that the term 'managed' is the key criterion in determining a communal establishment, even if the manager does not live on site full-time or at all. Continuity in this definition would aid comparisons over time, for example, in research on specialised communal populations such as older people. In addition, changing the definition of an establishment would impact on the size of the household population, which is used as a base or a denominator for a wide range of measures. Therefore it seems appropriate to keep the 2001 definition for 2011, subject to clarification of a few 'grey' areas.

Certain types of accommodation are particularly problematic when categorising residents as living in either a communal establishment or a household. These include sheltered accommodation units for older people and purpose-built student cluster flats. The increasing prevalence of these types of accommodation compounds the scale of the problem; for example the rise in student numbers in higher education has led to the construction of more non-traditional student accommodation.

Student accommodation can illustrate the types of difficulties encountered. Students in a traditional hall of residence offering meals and owned by their university are clearly part of the communal population. It is equally clear that groups of students who rent a house from a private landlord are part of the household population. However between these extremes lie a wide variety of other forms of student accommodation. For example, a university might own some large houses in a residential area that each accommodates several students. Or a private company may manage purpose-built flats in which students have their own rooms and bathrooms with shared kitchens. Are either of these communal establishments?

There are various criteria that can be used to define whether accommodation is communal or not:

- number of residents
- type of living arrangements (such as shared facilities)
- ownership of the property
- whether the accommodation is available to anybody or only a select group (such as older people or students)

Some of these criteria could be identified by enumerators on Census day while others would need to be determined in advance.

The working group recommended that:

'Accommodation available only to students (and not to anybody else) should be defined as communal. This would include university-owned cluster flats, university-owned houses and similar accommodation owned by a private company and provided solely for students. It would exclude houses rented to students by private landlords, as these are part of the general housing stock.'

Sheltered accommodation for older people is another hard to define area. Where an establishment provides all meals to its residents and they do not have separate cooking facilities, the accommodation is clearly communal. Conversely, where sheltered accommodation is provided in self-contained flats with their own kitchen facilities, the residents form part of the household population. However the status of accommodation between these extremes may be less clear. In 2001 a sheltered accommodation establishment was defined as communal if fewer than 50 per cent of units had their own cooking facilities; the working group recommended continued use of this cut-off to ensure consistency over time.

While many members of staff working in 'communals', and their families, will not live at the establishment, others may live on site. A criterion needs to be provided to distinguish between those included in the communal population and those living in private households within the establishment. In the latter case, the working group recommended that if they live in a household space that is a self-contained unit with its own door, then people are classified as part of the household population. There is a compelling argument for having such staff counted in private households where possible, since the census normally asks for information about relationships between household members, but not between people living in establishments. Hence family information would be lost if staff households were enumerated as part of the communal population.

Households

Having defined the household population (by specifying who is part of the communal population), the key remaining issue is how to divide up the household population into separate households. In most cases this is straightforward. A family or an individual living alone in a dwelling is generally thought of as a household. However, should a lodger living in their spare room be part of their household or in a separate one-person household?

The main difficulty in defining households arises where unrelated adults share a dwelling. For example if five students are sharing a privately-rented dwelling, does this represent one multi-person household or five separate one-person households? Most working group members stated that they would prefer the former, but others recognised the need for flexibility. Any decision made about this will impact upon the number of households estimated by the census.

A household can conceptually be defined in a variety of ways, including:

- by the relationships between people living in a dwelling (are they family, friends or strangers who live in the same accommodation?)
- by shared activities (for example, paying bills, eating together)
- by the facilities shared (for example, living room, bathroom or kitchen).

In the 2001 Census, the household definition included aspects of the three concepts above:

'A household is
- one person living alone, or
- a group of people (not necessarily related) living at the same address with common housekeeping – sharing either a living room or sitting room, or at least one meal a day.'

Users believe that the 2001 definition is a good starting point for 2011 and that consistency over time is important. However they argue that the term 'common housekeeping' is outdated and would be poorly understood in 2011. In addition, the definition needs to reflect social changes and in this context the clause referring to 'sharing at least one meal a day' was thought problematic, given that many families do not sit down and eat together regularly.

The working group recommended the following working definition:

'A household is
- one person living alone, or
- a group of people (not necessarily related) living at the same address and sharing cooking facilities *and* some living space.'

Whereas the 2001 definition required either a social element (sharing meals) or an accommodation element (sharing a living or sitting room), this recommended definition has moved to a solely accommodation-based household definition. This would provide a more objective basis with which to define a household.

DISCUSSION

The aim of the report *A Demographic Statistics Service for the 21st Century* was to stimulate practical thinking to nurture a more flexible series of output bases. These are needed to measure the increasingly more mobile population and more diverse living arrangements in modern society. The key requirement for users remains accurate delivery of outputted data on a usual residence base; however, the flexibility afforded by other population bases at output is important for other purposes such as performance monitoring.

Comparisons between the output bases available from the 2001 Census and those theoretically available from the 2011 Census, subject to implementation of the recommended definitions discussed above, would indicate the feasibility of moving towards this goal of greater flexibility. Table 2 assesses potential improvements over the 2001 provision, and merits more detailed discussion.

| Table 2 | | Population bases potentially available from the 2011 Census and other sources |

Output Base[1]	Potentially available from 2011 Census?	Examples of alternative sources in Census year and intercensal years
1. **Usual residence** – the population at the address at which they *usually* live	Yes – planned	Mid-year population estimates (based on census usually resident population; slightly different definition of usual residence)
2. **Household population** – the population living in private households	Yes – planned	Mid-year household population estimates (based on census usually resident population and census proportion in households)
3. **Communal population** – the population living in communal establishments	Yes – planned	Data available for some types of establishment, for example, prisoners data from Home Office; student data from HESA
4. **Out of term population** – the population usually resident in an area out of term-time	Yes (possible; would need to ask students and boarding school pupils for their vacation address, if different)	
5. **Seasonal population** – either the usually resident population at different times of year or enhanced to include visitors (for example, tourists)	No	Experimental quarterly population estimates (based on census usually resident population) estimate the resident population for each quarter
6. *De facto* **population** – population present on Census night	Unlikely in 2011 (though possible if visitor data is of high quality and resources available to match visitors back to their usual residence and remove them from the population there)	
7. **Legal population** – any 'legal' definition of the population to serve a particular purpose	No	
8. **Week-day population** – population usually resident/present overnight during the week (for example, Monday to Thursday nights)	Possible. Could be estimated for 2011 if secondary residence data available (address plus purpose or frequency of stay).	
9. **Weekend population** – population usually resident/present at the weekend (for example, Friday night to Sunday)	Possible. Similar to usual residence under proposed Census definition if Census day is on a weekend; quality of estimate for 2011 would be improved if secondary residence data available (address plus purpose or frequency of stay)	
10. **Temporarily resident population** – those present in an area but are not usually resident there (for example, divided children, temporary migrant workers)	Yes – planned. Available from visitor data but accuracy of this base would be dependent on the quality of visitor information collected in 2011	Some data on sub-groups from for example, International Passenger Survey, British Tourism Survey
11. **Work-day population** – population present in an area during normal working hours Monday to Friday	Possible. Could be partly estimated from workplace data. Place of study data also needed for the full picture	Survey data on employment. Data on pupils/students from, for example, DfES or educational institutions
12. **Average population** – average of various other population bases	Possible; dependent upon other output bases available	
13. **UK residents living abroad** – temporarily or for part of the time (for example, international commuters)	No. But secondary residence data would provide some information on this, if collected in 2011	Some information may be available from, for example, electoral registers, health or National Insurance sources. Potential to obtain limited data from household surveys or from International Passenger Survey
14. **Non-UK residents living in the UK** – for example, visitors, short-term migrants, international commuters	Yes – planned. Available from visitor data but accuracy of this base would be dependent on the quality of visitor information collected in 2011	International Passenger Survey (visitors); National Insurance data, work permits (temporary foreign workers)
15. **Bespoke service populations** – target population for a specific service/policy	Dependent upon nature of output needed	Health or administrative sources

1   As described in *A Demographic Statistics Service for the 21st Century*[1].

Many of the output bases in Table 2 could be fully or partly provided by the 2011 Census, though the extent would be determined by the range of variables collected. The three output bases available in 2001 would be retained in 2011 (that is, usual residence, household population and communal population).

In addition, a further four output bases in Table 2 could potentially be available in 2011 if the working group's recommendations were followed. These would be:

- Week-day population; this could be estimated if secondary residence data were available (address plus purpose or frequency of stay). However, the accuracy of this estimation would be determined by the type of questions asked
- Weekend population; this might be approximately equivalent to usual residence as a result of a tightened definition of usual residence; the quality of estimation would be improved if secondary residence data were available (address plus purpose or frequency of stay)
- Temporarily resident population; this would be obtained from visitor data in 2011

- Out of term population; this would be feasible if students and boarding school pupils were asked for their vacation address, as in 2001

A further benefit might be increased flexibility within the bespoke service populations, affording greater opportunities for outputs based especially upon visitor data. This would constitute a potential improvement on the 2001 bespoke outputs.

Some output bases not available from the census could be derived partially from administrative sources or surveys, but such sources are often limited in coverage and would require further refinement before being considered suitable as a population base.

CONCLUSIONS

The list of possible output bases listed by the 2003 report is not necessarily exhaustive; other potential bases may emerge in continuing discussions within the working group. For example, producers and users of labour market statistics have expressed a need for an output base combining usual residents with temporary workers from abroad to produce a more comprehensive population base for employment statistics. Delivery of this and similar output bases would be determined by both technical feasibility and resources.

On balance, significant improvements to the flexibility of outputs could more readily be achieved if these recommended definitions were applied to the enumeration base of usual residents plus visitors in 2011. The implications of this for authorities with large numbers of people living there for only part of the time, for example during the working week, could be very significant. Similarly, weekend populations might be better measured as a result of a more comprehensive working definition of usual residence. The temporarily resident population on Census night would give an indication of numbers of foreign and domestic visitors, which would be valuable for local areas. All of these would represent an improvement over what was possible for 2001, providing a step towards a more flexible data output platform for the 21st century, consistent with the aims of the 2003 report.

As indicated in the background section, the 2011 Census will be limited by questionnaire space and a requirement to keep the enumeration process as simple as possible. Accordingly, the next step is for ONS to review the strength of these cases and give appropriate priority to each one, following discussions with users. Development work is also underway within ONS to investigate the quality of information that can be collected from a self-completion question on the topics that are necessary to deliver flexible population outputs from the 2011 Census.

ACKNOWLEDGEMENTS

The authors acknowledge the invaluable support of members of the Population Definitions Working Group in discussing these issues. Thanks are also due to colleagues within ONS for comments on early drafts of this article, and especially to Roma Chappell and Peter Goldblatt for their encouragement and constructive advice.

REFERENCES

1   Office for National Statistics (2003) *A Demographic Statistics Service for the 21st Century*. www.statistics.gov.uk/about/ methodology_by_theme/Dem_Stat_Ser_21ST_Cen.asp

2   Baker L (2004) Covering All Bases: early thoughts for population bases for the 2011 Census. *Population Trends* **116**, pp 6–10. www.statistics.gov.uk/StatBase/Product.asp?vlnk=6303&Pos=&Col Rank=1&Rank=422

3   Office for National Statistics, General Register Office Scotland and Northern Ireland Statistics and Research Agency (2004) *Census 2001: Definitions*. TSO: London. www.statistics.gov.uk/statbase/ Product.asp?vlnk=12951

4   United Nations Economic Commission for Europe and the Statistical Office of the European Communities (2006) *Draft Conference of European Statisticians recommendations for the 2010 Censuses of Population and Housing*.

5   Office for National Statistics (2006) Information Paper – *The 2011 Census: Assessment of initial user requirements on content for England and Wales*. www.statistics.gov.uk/about/consultations/ 2011Census_response.asp

6   Blackwell L, Lynch K, Smith J and Goldblatt P (2003) *Longitudinal Study 1971–2001: Completeness of Census Linkage*. Series LS no.10. Office for National Statistics: London. www.statistics.gov.uk/ StatBase/Product.asp?vlnk-10721&Pos=&ColRank=1&Rank=272

7   Office for National Statistics (2003) 2001 Census data (S001 and S126).

Tables

* Numbers in brackets indicate former table numbers in editions of *Population Trends* prior to spring 1999 (No 95). Former tables 16 and 17 (Deaths by selected causes, and Abortions) now appear in *Health Statistics Quarterly*.

Population Trends tables are also available in XLS or CSV formats via our website www.statistics.gov.uk

Symbols
.. not available   – nil or less than half the final digit shown
: not applicable   p provisional

Table 1.1 — Population and vital rates: international

Selected countries   Numbers (thousands)/Rates per thousand

Year	United Kingdom	Austria	Belgium	Cyprus[1,3]	Czech Republic[3]	Denmark	Estonia[3]	Finland	France	Germany[2]	Greece	Hungary[3]	Irish Republic
Population (thousands)													
1971	55,928	7,501	9,673	..	9,810	4,963	1,369	4,612	51,251	78,313	8,831	10,370	2,992
1976	56,216	7,566	9,818	498	10,094	5,073	1,435	4,726	52,909	78,337	9,167	10,590	3,238
1981	56,357	7,569	9,859	515	10,293	5,121	1,482	4,800	54,182	78,408	9,729	10,712	3,443
1986	56,684	7,588	9,862	545	10,340	5,120	1,534	4,918	55,547	77,720	9,967	10,631	3,543
1991	57,439	7,813	9,979	587	10,309	5,154	1,566	5,014	57,055	79,984	10,247	10,346	3,526
1996	58,164	7,953	10,160	750[10]	10,321	5,260	1,470	5,117	58,030	81,900	10,480	10,190	3,630
1997	58,314	7,965	10,180	760[10]	10,300	5,275	1,460	5,140	58,610	82,060	10,500	10,150	3,660
1998	58,475	7,980	10,200	770[10]	10,290	5,295	1,450	5,147	58,400	82,030	10,520	10,110	3,700
1999	58,684	7,990	10,230	770[10]	10,280	5,330	1,440	5,170	58,620	82,060	10,530	10,070	3,740
2000	58,886	8,010	10,250	780[10]	10,270	5,340	1,372	5,180	58,900	82,180	10,010	10,020	3,790
2001	59,113	8,040	10,290	790[10]	10,220	5,360	1,360	5,190	59,190	82,350	10,020	10,190	3,840
2002	59,322	8,080	10,330	800[10]	10,206	5,370	1,361	5,200	59,490	82,490	10,988	10,160	3,920
2003	59,554	8,120	10,380	810[10]	10,203	5,390	1,350	5,210	59,770	82,530	11,018	10,130	3,980
2004	59,834[11]	8,170	10,396	..	10,212	5,400	1,351[P]	5,230	60,200	82,500	11,041	10,117	4,028
2005	..	8,206[13p]	10,446[13p]	..	10,221[13p]	5,411[13p]	1,347[13p]	5,237[13p]	60,561[13p]	82,501[13p]	11,076[13p]	10,097[13p]	4,109[13p]
Population changes (per 1,000 per annum)													
1971–76	1.0	1.7	3.0	..	5.8	4.4	9.6	4.9	6.5	0.1	7.6	4.2	16.4
1976–81	0.5	0.1	0.8	6.8	3.9	1.9	6.6	3.1	4.8	0.2	12.3	2.3	12.7
1981–86	1.2	0.5	0.1	11.7	0.9	0.0	7.0	4.9	5.0	−1.8	4.9	−1.5	5.8
1986–91	2.7	5.9	2.4	15.4	−0.6	1.3	4.2	3.9	5.4	5.8	5.6	−5.4	−1.0
1991–96	2.5	3.6	3.6	55.5	0.1	4.1	−12.4	3.8	3.4	4.8	4.5	−3.0	4.3
1997–98	2.8	1.9	2.0	13.2	−1.0	3.8	−6.8	1.4	−3.6	−0.4	1.9	−3.9	10.9
1998–99	3.6	1.3	2.9	10.0	−1.0	6.6	−16.9	4.5	3.8	−0.4	1.0	−4.0	10.8
1999–2000	3.4	2.5	2.0	13.0	−1.0	1.9	−47.2	1.9	4.8	1.5	−49.4	−5.0	13.4
2000–01	3.9	3.7	3.9	12.8	−4.9	3.7	−8.7	1.9	4.9	2.1	1.0	17.0	13.2
2001–02	3.5	5.0	3.9	12.7	−2.0	1.9	0.0	1.9	5.1	1.7	96.8	−2.9	20.8
2002–03	3.9	5.0	4.8	12.5	0.0	3.7	−7.4	1.9	4.7	0.5	2.7	−3.0	15.3
2003–04	4.7	6.2	1.5	..	1.2	1.9	0.7	3.8	7.2	−0.4	1.9	−1.3	12.1
2004–05	..	4.4	4.8	..	0.9	2.0	−3.0	1.3	6.0	0.0	3.2	−2.0	20.1
Live birth rate (per 1,000 per annum)													
1971–75	14.1	13.3	13.4	17.7	17.8	14.6	15.4	13.1	16.0	10.5	15.8	16.1	22.2
1976–80	12.5	11.5	12.5	19.0	17.1	12.0	15.0	13.6	14.1	10.5	15.6	15.8	21.3
1981–85	12.9	12.0	12.0	20.2	13.5	10.2	15.6	13.4	14.2	10.7	13.3	12.3	19.2
1986–90	13.7	11.6	12.1	18.8	12.7	11.5	15.5	12.7	13.8	9.8	10.6	11.8	15.8
1991–95	13.2	11.8	12.0	16.9	11.1	13.1	10.7	12.9	12.7	10.9	9.9	11.7	14.0
1996	12.6	11.0	11.5	14.5	8.8	12.9	9.0	11.8	12.6	9.7	9.6	10.3	13.9
1997	12.5	10.4	11.4	13.9	8.8	12.8	8.7	11.5	12.4	9.9	9.7	9.9	14.4
1998	12.3	10.1	11.2	13.1	8.8	12.5	8.4	11.1	12.6	9.7	9.6	9.6	14.5
1999	11.9	9.7	11.1	12.4	8.7	12.4	8.7	11.1	12.6	9.4	11.0	9.4	14.2
2000	11.5	9.7	11.2	12.2	8.8	12.6	9.6	11.0	13.2	9.3	11.7	9.7	14.3
2001	11.3	9.3	11.1	11.6	8.9	12.2	9.3	10.8	13.1	9.0	10.2	9.7	15.1
2002	11.3	9.7	..	11.1	9.1	11.9	9.6	10.7	12.8	8.7	..	9.5	15.5
2003	11.7	9.4	..	..	9.2	12.0	..	10.9	12.7	8.6	9.5	9.3	15.7
2004	12.1[12]	9.7	11.2	..	9.6	11.9	10.4	11.4	..	8.6	..	..	15.3
Death rate (per 1,000 per annum)													
1971–75	11.8	12.6	12.1	9.9	12.4	10.1	11.1	9.5	10.7	12.3	8.6	11.9	11.0
1976–80	11.9	12.3	11.6	10.4	12.5	10.5	12.1	9.3	10.2	12.2	8.8	12.9	10.2
1981–85	11.7	12.0	11.4	10.0	12.8	11.1	12.3	9.3	10.1	12.0	9.0	13.7	9.4
1986–90	11.4	11.1	10.8	10.2	12.4	11.5	11.9	9.8	9.5	11.6	9.3	13.5	9.1
1991–95	11.1	10.4	10.4	9.0	11.6	11.9	13.9	9.8	9.1	10.8	9.5	14.3	8.8
1996	10.9	10.0	10.3	8.5	10.9	11.6	12.9	9.6	9.2	10.8	9.6	14.0	8.7
1997	10.8	9.8	10.2	8.8	10.9	11.3	12.7	9.6	9.0	10.5	9.5	13.7	8.6
1998	10.8	9.7	10.3	8.0	10.6	11.0	13.4	9.6	9.2	10.4	9.8	13.9	8.5
1999	10.8	9.7	10.3	7.4	10.7	11.1	12.8	9.5	9.2	10.4	9.9	14.2	8.5
2000	10.3	9.5	10.2	7.7	10.6	10.9	13.4	9.5	9.1	10.2	10.5	13.5	8.2
2001	10.2	9.2	10.1	6.9	10.5	10.9	13.6	9.4	9.0	10.0	10.2	13.0	7.8
2002	10.2	9.5	..	7.3	10.6	10.9	13.5	9.5	9.0	10.2	..	13.1	7.5
2003	10.3	9.5	..	..	10.9	10.7	..	9.4	9.2	10.3	9.6	13.4	7.4
2004	9.7[12]	9.1	9.8	..	10.5	10.3	13.2	9.1	..	10.0	..	..	7.0
2005[p]	9.7[14]	..	..	..	..	..	..	..	..	..	..	..	..

Note:
Estimated population (mid-year), live birth and death rates up to the latest available date, as given in the *United Nations Monthly Bulletin of Statistics (November 2005)*, the *United Nations Demographic Yearbook (2000 Edn)*, *Eurostat Yearbook 2004* and the *Eurostat website (April 2006)*.

1 Republic of Cyprus - Greek Cypriot controlled area only
2 Including former GDR throughout.
3 The European Union consists of 25 member countries (EU25) - 1 May 2004 (10 new member countries).
4 Including the Indian held part of Jammu and Kashmir, the final status of which has not yet been determined.
5 Rates are based on births to or deaths of Japanese nationals only.
6 Excludes Hong Kong.
7 Estimate prepared by the Population Division of the United Nations.
8 Includes Hong Kong.
9 Rate is for 1990–1995.
10 Indicates population estimates of uncertain reliability.
11 Data for mid-2004 for the United Kingdom, Great Britain, England and Wales and England were revised due to the Harrow correction that was published on 20 December 2005.
12 Birth and death rates for 2004 have been calculated using the revised mid-2004 population estimates published on 20 December 2005.
13 As at 1 January.
14 Death rates for 2005 are based on the 2004 based Population Projections for 2005.
p provisional

Table 1.1 continued **Population and vital rates: international**

Selected countries Numbers (thousands)/Rates per thousand

Year	United Kingdom	Italy	Latvia[3]	Lithuania[3]	Luxembourg	Malta[3]	Netherlands	Poland[3]	Portugal	Slovakia[3]	Slovenia[3]	Spain	Sweden	EU–25[3]
Population (thousands)														
1971	55,928	54,073	2,366	3,160	342	330	13,194	32,800	8,644	4,540	1,732	34,216	8,098	..
1976	56,216	55,718	2,465	3,315	361	330	13,774	34,360	9,356	4,764	1,809	36,118	8,222	420,258
1981	56,357	56,502	2,515	3,422	365	322	14,247	35,902	9,851	4,996	1,910	37,741	8,320	428,563
1986	56,684	56,596	2,588	3,560	368	344	14,572	37,456	10,011	5,179	1,975	38,536	8,370	433,555
1991	57,439	56,751	2,662	3,742	387	358	15,070	38,245	9,871	5,283	2,002	38,920	8,617	440,927
1996	58,164	57,380	2,460	3,615	420	380	15,530	38,620	10,060	5,368	1,990	39,430	8,838	447,522
1997	58,314	57,520	2,430	3,580	416	380	15,610	38,650	10,090	5,379	1,987	39,520	8,845	448,785
1998	58,475	57,590	2,410	3,550	430	390	15,710	38,670	10,130	5,388	1,985	39,650	8,848	449,121
1999	58,684	57,650	2,390	3,520	427	390	15,810	38,650	10,170	5,393	1,978	39,840	8,860	449,994
2000	58,886	57,760	2,370	3,512	440	389	15,910	38,260	10,230	5,399	1,988	40,170	8,870	450,287
2001	59,113	57,950	2,364	3,480	440	390	16,050	38,250	10,290	5,379	1,990	40,610	8,900	452,043
2002	59,322	57,160	2,340	3,470	450	390	16,150	38,230	10,370	5,379	2,000	41,200	8,920	453,772
2003	59,554	57,610	2,332	3,460	450	400	16,220	38,200	10,440	5,379	2,000	41,870 [P]	8,960	455,764
2004	59,834 [11]	58,170	2,310	3,440	450	400	16,270	38,180	10,500 [P]	5,380	2,000	42,345	8,990	457,645
2005	..	58,462 [13p]	2,306 [13p]	3,425 [13p]	455 [13p]	403 [13p]	16,305 [13p]	38,174 [13p]	10,529 [13p]	5,385 [13p]	1,998 [13p]	43,038 [13p]	9,011 [13p]	..
Population changes (per 1,000 per annum)														
1971–76	1.0	6.1	8.4	9.8	10.7	0.0	8.8	9.5	16.5	9.9	8.9	11.1	3.1	..
1976–81	0.5	2.8	4.1	6.5	2.5	−4.8	6.9	9.0	10.6	9.7	11.2	9.0	2.4	4.0
1981–86	1.2	0.3	5.8	8.1	1.8	13.7	4.6	8.7	3.2	7.3	6.8	4.2	1.2	2.3
1986–91	2.7	0.5	5.7	10.2	10.2	8.1	6.8	4.2	−2.8	4.0	2.7	2.0	5.9	3.4
1991–96	2.5	2.2	−12.8	−1.7	17.0	8.4	4.6	2.0	3.8	3.4	−1.1	2.6	1.2	3.0
1997–98	2.8	1.2	−8.2	−8.4	33.7	26.3	6.4	0.5	4.0	1.5	−1.0	3.3	0.3	0.7
1998–99	3.6	1.0	−8.3	−8.5	−0.7	0.0	6.4	−0.5	3.9	0.7	−3.5	4.8	1.4	1.9
1999–2000	3.4	1.9	−8.4	−2.3	30.4	−2.6	6.3	−10.1	5.9	0.9	5.1	8.3	1.1	0.7
2000–01	3.9	3.3	−4.2	−9.1	0.0	2.6	8.8	−0.3	5.9	−3.7	1.0	11.0	3.4	3.9
2001–02	3.5	−13.6	−8.5	−2.9	22.7	0.0	6.2	−0.5	−7.8	0.0	5.0	14.5	2.2	3.8
2002–03	3.9	7.9	−4.3	−2.9	0.0	25.6	4.3	−0.8	6.8	0.0	0.0	16.3	4.5	4.4
2003–04	4.7	9.7	−8.6	−5.8	0.0	0.0	3.1	−0.5	5.7	0.0	0.0	11.3	3.3	2.5
2004–05	..	5.0	−1.7	−4.4	11.1	7.5	2.2	−0.2	2.8	0.9	−1.0	16.4	2.3	..
Live birth rate (per 1,000 per annum)														
1971–75	14.1	16.0	14.4	16.4	11.6	17.5	14.9	17.9	20.3	19.7	16.4	19.2	13.5	..
1976–80	12.5	12.6	13.9	15.4	11.2	17.0	12.6	19.3	17.9	20.3	16.3	17.1	11.6	..
1981–85	12.9	10.6	15.2	16.0	11.6	15.3	12.2	19.0	14.5	18.0	14.2	12.8	11.3	..
1986–90	13.7	9.8	15.3	15.8	12.2	16.0	12.8	15.5	11.9	15.8	12.3	10.8	13.2	..
1991–95	13.2	9.6	10.8	13.1	13.3	14.0	12.8	12.9	11.4	13.3	10.0	9.8	13.3	..
1996	12.6	9.2	7.9	10.5	13.7	13.5	12.2	11.1	11.1	11.2	9.4	9.2	10.8	10.8
1997	12.5	9.4	7.6	10.2	13.1	13.1	12.3	10.7	11.4	11.0	9.1	9.4	10.2	10.7
1998	12.3	9.3	7.5	10.4	12.7	12.2	12.7	10.2	11.4	10.7	9.0	9.3	10.1	10.5
1999	11.9	9.3	8.0	10.3	13.0	11.4	12.7	9.9	11.6	10.4	8.8	9.6	10.0	10.5
2000	11.5	9.4	8.3	9.8	13.1	10.8	13.0	9.8	11.8	10.2	9.1	9.8	10.2	10.6
2001	11.3	9.3	8.3	9.1	12.4	..	12.6	9.5	10.8	9.5	8.8	..	10.3	10.4
2002	11.3	9.3	8.6	8.7	12.0	..	12.6	9.2	11.0	9.5	8.8	..	10.7	10.3
2003	11.7	9.4	..	8.9	11.5	..	12.4	9.2	11.2	9.7	8.7	..	11.1	..
2004	12.1 [12]	..	..	8.9	12.1	..	11.9	9.3	10.4	10.0	9.0	..	11.2	..
Death rate (per 1,000 per annum)														
1971–75	11.8	9.8	11.6	9.0	12.2	9.0	8.3	8.4	11.0	9.4	10.0	8.5	10.5	..
1976–80	11.9	9.7	12.6	10.1	11.5	9.0	8.1	9.2	10.1	9.8	9.8	8.0	10.9	..
1981–85	11.7	9.5	12.8	10.6	11.2	8.2	8.3	9.6	9.6	10.1	10.3	7.7	11.0	..
1986–90	11.4	9.4	12.4	10.3	10.5	7.4	8.5	10.0	9.6	10.1	9.6	8.2	11.1	..
1991–95	11.1	9.7	14.8	12.0	9.8	7.6	8.8	10.2	10.4	9.9	9.7	8.7	10.9	..
1996	10.9	9.6	13.8	11.6	9.4	7.4	8.9	10.0	10.8	9.8	9.4	8.9	10.6	10.1
1997	10.8	9.8	13.8	11.1	9.4	7.7	8.7	9.8	10.6	9.7	9.5	8.9	10.5	10.0
1998	10.8	10.0	14.2	11.5	9.2	8.1	8.8	9.7	10.7	9.9	9.6	9.2	10.5	10.0
1999	10.8	9.9	13.7	11.4	8.8	8.2	8.9	9.9	10.8	9.7	9.5	9.1	10.7	10.0
2000	10.3	9.7	13.2	11.1	8.6	7.6	8.8	9.5	10.6	9.8	9.3	9.1	10.5	9.8
2001	10.2	9.6	14.0	11.6	8.4	..	8.8	9.4	10.4	9.6	9.3	8.9	10.5	9.7
2002	10.2	9.7	13.9	11.8	8.4	..	8.9	9.4	10.2	9.6	9.3	..	10.7	9.8
2003	10.3	10.8	..	11.9	8.6	..	8.7	9.6	10.9	9.7	9.7	..	10.4	..
2004	9.7 [12]	..	..	12.0	7.9	..	8.4	..	9.7	9.6	9.3	..	10.1	..
2005[P]	9.7 [14]	..	..	..	..	..	..	..	..	..	..	..	..	..

See notes on first page of table.

Table 1.1 continued	Population and vital rates: international									

Selected countries

Numbers (thousands)/Rates per thousand

Year	United Kingdom	EU–25[3]	Russian Federation	Australia	Canada	New Zealand	China	India[4]	Japan[5]	USA
Population (thousands)										
1971	55,928	..	130,934	13,067	22,026	2,899	852,290 [6]	551,311	105,145	207,661
1976	56,216	420,258	135,027	14,033	23,517	3,163	937,170 [6]	617,248	113,094	218,035
1981	56,357	428,563	139,225	14,923	24,900	3,195	1,008,460 [6]	675,185	117,902	229,958
1986	56,684	433,555	144,154	16,018	26,204	3,317	1,086,733 [6]	767,199	121,672	240,680
1991	57,439	440,927	148,245	17,284	28,031	3,477	1,170,100 [6]	851,897	123,964	252,639
1996	58,164	447,552	147,739	18,311	29,610	3,730	1,223,890 [6,10]	941,580 [10]	125,761	265,463
1997	58,314	448,785	147,105	18,524	29,910	3,780	1,236,260 [6,10]	959,800 [10]	126,065	268,008
1998	58,475	449,121	146,540	18,710	30,160	3,820	1,248,100 [6,10]	978,080 [10]	126,400	270,300
1999	58,684	449,994	145,940	18,930	30,490	3,840	1,259,090 [6,10]	996,430 [10]	126,630	272,691
2000	58,886	450,287	145,560	19,150	30,770	3,860	1,275,130 [7,8,10]	1,014,820 [10]	126,840	275,260
2001	59,113	452,043	145,980	19,410	31,110	3,850	..	1,033,325 [10]	127,130	284,800
2002	59,322	453,772	145,310	19,640	31,410	3,940	..	1,050,640 [10]	127,400	288,370 [7]
2003	59,554	455,764	144,570	19,870	31,630	4,010	..	1,068.210 [10]	127,650	290,810
2004	59,834 [11]	457,645	143,820	20,110	31,950	4,060	..	1,085,600 [10]	126,670	291,685
2005	..	..	..	..	..	..	..	..	..	..
Population changes (per 1,000 per annum)										
1971–76	1.0	..	6.3	14.8	13.5	18.2	19.9 [6]	23.9	15.1	10.0
1976–81	0.5	4.0	6.2	12.7	11.8	2.0	15.2 [6]	18.8	8.5	10.9
1981–86	1.2	2.3	7.1	14.7	10.5	7.6	15.5 [6]	27.3	6.4	9.3
1986–91	2.6	3.4	5.7	15.8	13.9	9.6	15.3 [6]	22.1	3.8	9.9
1991–96	2.5	3.0	–0.7	11.9	11.3	15.1	9.2 [6]	21.1	2.9	10.2
1997–98	2.8	0.7	–3.8	10.0	8.4	10.6	9.6 [6]	19.0	2.7	8.6
1998–99	3.6	1.9	–4.1	11.8	10.9	5.2	8.8 [6]	18.8	1.8	8.8
1999–2000	3.4	0.7	–2.6	11.6	9.2	5.2	12.7 [8]	18.5	1.7	9.4
2000–01	3.9	3.9	2.9	13.6	11.0	–2.6	..	18.2 [P]	2.3	34.7
2001–02	3.5	3.8	–4.6	11.8	9.6	23.4	..	16.8 [P]	2.1	12.5
2002–03	3.9	4.4	–5.1	11.7	7.0	17.8 [P]	..	16.7 [P]	2.0	8.5
2003–04	4.7	2.5	–5.2	12.1	10.1	12.5	..	16.3 [P]	0.2	3.0
Live birth rate (per 1,000 per annum)										
1971–75	14.1	..	..	18.8	15.9	20.4	27.2 [6]	35.6	18.6	15.3
1976–80	12.5	..	..	15.7	15.5	16.8	18.6 [6]	33.4	14.9	15.2
1981–85	12.9	..	..	15.6	15.1	15.8	19.2 [6]	..	12.6	15.7
1986–90	13.7	..	..	15.1	14.8	17.1	..	..	10.6	16.0
1991–95	13.2	..	10.2	..	..	..	18.5 [6,9]	..	..	..
1996	12.6	10.8	8.8	13.9	12.3	15.4	9.8 [6]	27.3	9.6	14.7
1997	12.5	10.7	8.6	13.6	11.6	15.4	9.1 [8]	..	9.5	14.5
1998	12.3	10.5	8.7	13.3	11.3	14.5	8.1 [8]	26.2	9.5	14.6
1999	11.9	10.5	8.3	13.1	11.0	14.9	7.8 [8]	..	9.3	14.5
2000	11.5	10.6	8.6	13.0	10.8	14.7	8.1 [8]	..	9.4	14.7
2001	11.3	10.4	9.0	12.7	..	14.4	7.2 [8]	..	9.2	14.1
2002	11.3	10.3	9.6	12.8	..	13.7	7.1 [8]	..	9.1	13.9
2003	11.7	..	10.2	12.6	..	14.0	6.9 [8]	..	8.8	14.1
2004	12.1 [12]	..	10.5	12.7	..	14.3	7.2 [8]	..	..	..
Death rate (per 1,000 per annum)										
1971–75	11.8	..	..	8.2	7.4	8.4	7.3 [6]	15.5	6.4	9.1
1976–80	11.9	..	..	7.6	7.2	8.2	6.6 [6]	13.8	6.1	8.7
1981–85	11.7	..	..	7.3	7.0	8.1	6.7 [6]	..	6.1	8.6
1986–90	11.4	..	..	7.2	7.3	8.2	..	..	6.4	8.7
1991–95	11.1	..	13.7	..	..	..	..	..	..	..
1996	10.9	10.1	14.1	7.0	7.2	7.6	5.0 [6]	8.9	7.1	8.7
1997	10.8	10.0	13.7	7.0	7.2	7.3	4.9 [8]	..	7.2	8.6
1998	10.8	10.0	13.6	6.8	7.2	6.9	5.0 [8]	9.0	7.4	8.6
1999	10.8	10.0	14.7	6.8	7.4	7.3	5.0 [8]	..	7.8	8.8
2000	10.3	9.8	15.3	6.7	7.5	6.9	5.1 [8]	..	7.6	8.7
2001	10.2	9.7	15.6	6.6	..	7.2	5.0 [8]	..	7.6	8.5
2002	10.2	9.8	..	6.8	..	7.1	5.0 [8]	..	7.7	8.5
2003	10.3	..	..	6.6	..	7.0	5.4 [8]	..	7.9	8.4
2004	9.7 [12]	..	..	6.6	..	7.0	5.3 [8]	..	..	..
2005[P]	9.7 [14]	..	..	..	..	..	..	..	..	..

See notes on first page of table.

Table 1.2	**Population: national**

Constituent countries of the United Kingdom

Numbers (thousands) and percentage age distribution

Mid-year	United Kingdom	Great Britain	England and Wales	England	Wales	Scotland	Northern Ireland
Estimates							
1971	55,928	54,388	49,152	46,412	2,740	5,236	1,540
1976	56,216	54,693	49,459	46,660	2,799	5,233	1,524
1981	56,357	54,815	49,634	46,821	2,813	5,180	1,543
1986	56,684	55,110	49,999	47,188	2,811	5,112	1,574
1991	57,439	55,831	50,748	47,875	2,873	5,083	1,607
1993[3]	57,714	56,078	50,986	48,102	2,884	5,092	1,636
1994[3]	57,862	56,218	51,116	48,229	2,887	5,102	1,644
1995[3]	58,025	56,376	51,272	48,383	2,889	5,104	1,649
1996[3]	58,164	56,503	51,410	48,519	2,891	5,092	1,662
1997[3]	58,314	56,643	51,560	48,665	2,895	5,083	1,671
1998[3]	58,475	56,797	51,720	48,821	2,900	5,077	1,678
1999[3]	58,684	57,005	51,933	49,033	2,901	5,072	1,679
2000[3]	58,886	57,203	52,140	49,233	2,907	5,063	1,683
2001[3]	59,113	57,424	52,360	49,450	2,910	5,064	1,689
2002[3]	59,322	57,625	52,570	49,647	2,923	5,055	1,697
2003	59,554	57,851	52,794	49,856	2,938	5,057	1,703
2004[4]	59,834	58,124	53,046	50,093	2,952	5,078	1,710
2005	..	..	..	..	..	5,059	..
of which **(percentages)**							
0–4	*5.7*	*5.6*	*5.7*	*5.7*	*5.4*	*5.2*	*6.4*
5–15	*13.8*	*13.7*	*13.8*	*13.8*	*14.0*	*13.0*	*16.0*
16–44	*40.2*	*40.2*	*40.2*	*40.4*	*37.5*	*39.6*	*41.4*
45–64M/59F	*21.7*	*21.8*	*21.7*	*21.6*	*22.7*	*23.0*	*20.1*
65M/60F–74	*11.0*	*11.0*	*11.0*	*10.9*	*12.0*	*11.7*	*9.9*
75 and over	*7.6*	*7.6*	*7.7*	*7.6*	*8.4*	*7.4*	*6.2*
Projections[1]							
2006	60,533	58,800	53,691	50,714	2,977	5,108	1,733
2011	61,892	60,124	55,005	51,967	3,037	5,120	1,767
2016	63,304	61,504	56,378	53,276	3,102	5,126	1,800
2021	64,727	62,897	57,770	54,605	3,165	5,127	1,830
of which **(percentages)**							
0–4	*5.6*	*5.6*	*5.6*	*5.6*	*5.3*	*5.0*	*5.9*
5–15	*12.0*	*12.0*	*12.1*	*12.1*	*11.8*	*11.2*	*13.0*
16–44	*36.8*	*36.8*	*36.9*	*37.1*	*34.7*	*35.0*	*37.2*
45–64[2]	*25.9*	*25.9*	*25.8*	*25.8*	*25.9*	*27.6*	*25.6*
65–74[2]	*10.2*	*10.2*	*10.1*	*10.0*	*11.6*	*11.3*	*9.5*
75 and over	*9.5*	*9.5*	*9.5*	*9.4*	*10.6*	*10.0*	*8.6*

Note: Figures may not add exactly due to rounding.

1 National projections based on mid-2004 population estimates.

2 Between 2010 and 2020, state retirement age will change from 65 years for men and 60 years for women to 65 years for both sexes.

3 These revised population estimates were published on 9 September 2004 (for mid-2001 and mid-2002) and 7 October 2004 (for mid-1992 to mid-2000), following the local authority population studies, and replace all earlier versions. All figures shown on this table are now therefore on a consistent basis.

4 Data for mid-2004 for the United Kingdom, Great Britain, England and Wales and England were revised due to the Harrow correction that was published on 20 December 2005.

Tel no. for all queries relating to population estimates - 01329 813318

Table 1.3 — Population: subnational

Government Office Regions of England[1] — Numbers (thousands) and percentage age distribution

Mid-year	North East	North West	Yorkshire and The Humber	East Midlands	West Midlands	East	London	South East	South West
Estimates									
1971	2,679	7,108	4,902	3,652	5,146	4,454	7,529	6,830	4,112
1976	2,671	7,043	4,924	3,774	5,178	4,672	7,089	7,029	4,280
1981	2,636	6,940	4,918	3,853	5,187	4,854	6,806	7,245	4,381
1986	2,594	6,833	4,884	3,908	5,180	4,999	6,774	7,468	4,548
1991	2,587	6,843	4,936	4,011	5,230	5,121	6,829	7,629	4,688
1993[4]	2,594	6,847	4,954	4,056	5,246	5,154	6,844	7,673	4,734
1994[4]	2,589	6,839	4,960	4,072	5,249	5,178	6,874	7,712	4,757
1995[4]	2,583	6,828	4,961	4,092	5,257	5,206	6,913	7,763	4,782
1996[4]	2,576	6,810	4,961	4,108	5,263	5,233	6,974	7,800	4,793
1997[4]	2,568	6,794	4,958	4,120	5,262	5,267	7,015	7,853	4,827
1998[4]	2,561	6,792	4,958	4,133	5,271	5,302	7,065	7,889	4,849
1999[4]	2,550	6,773	4,956	4,152	5,272	5,339	7,154	7,955	4,881
2000[4]	2,543	6,774	4,959	4,168	5,270	5,375	7,237	7,991	4,917
2001[4]	2,540	6,773	4,977	4,190	5,281	5,400	7,322	8,023	4,943
2002[4]	2,538	6,783	4,993	4,223	5,304	5,422	7,371	8,044	4,968
2003	2,539	6,805	5,009	4,252	5,320	5,463	7,388	8,080	4,999
2004	2,545	6,827	5,039	4,280	5,334	5,491	7,429[5]	8,110	5,038
of which **(percentages)**									
0–4	5.3	5.6	5.6	5.5	5.8	5.7	6.5	5.6	5.1
5–15	13.6	14.2	14.0	13.9	14.3	13.9	12.9	13.9	13.3
16–44	39.1	39.5	39.7	39.3	39.3	38.6	48.7	39.1	36.9
45–64M/59F	22.6	22.0	21.9	22.5	21.8	22.4	18.0	22.3	22.9
65M/60F–74	11.8	11.3	11.1	11.2	11.3	11.4	8.2	11.0	12.3
75 and over	7.7	7.5	7.6	7.7	7.6	8.0	5.7	8.1	9.4
Projections[2]									
2004	2,535	6,811	5,022	4,275	5,330	5,499	7,431	8,122	5,031
2008	2,525	6,852	5,079	4,366	5,380	5,646	7,614	8,300	5,163
2013	2,516	6,914	5,154	4,479	5,451	5,833	7,858	8,527	5,328
2018	2,510	6,987	5,234	4,594	5,531	6,025	8,105	8,765	5,498
2023	2,502	7,057	5,313	4,706	5,609	6,212	8,331	9,005	5,668
2028	2,489	7,107	5,379	4,804	5,672	6,380	8,523	9,222	5,823
of which **(percentages)**									
0–4	4.7	5.3	5.4	5.1	5.6	5.4	6.3	5.4	4.8
5–15	11.2	12.1	12.1	11.9	12.5	12.3	11.7	12.1	11.3
16–44	33.7	35.4	35.7	33.9	34.7	34.0	43.8	35.0	32.8
45–64[3]	25.3	24.9	24.6	25.6	25.1	25.0	24.2	25.1	25.3
65–74[3]	12.7	11.1	11.0	11.5	10.7	11.2	7.4	10.8	12.1
75 and over	12.3	11.2	11.3	12.1	11.4	12.2	6.6	11.7	13.7

Note: Figures may not add exactly due to rounding.

1 From 1 April 2002 there are four Directorates of Health and Social Care (DHSCs) within the Department of Health. The GORs sit within the DHSCs as follows: North East, North West, Yorkshire and The Humber GORs are within North DHSC, East Midlands, West Midlands and East GORs are within Midlands and Eastern DHSC, London GOR equates to London DHSC and South East and South West GORs are within South DHSC. See 'In brief' *Health Statistics Quarterly* 15 for further details of changes to Health Areas.

2 These projections are based on the mid-2003 population estimates and are consistent with the 2003-based national projections produced by the Government Actuary's Department and presented in Table 1.2 in *Health Statistics Quarterly* nos 25 to 28.

3 Between 2010 and 2020, state retirement age will change from 65 years for men and 60 years for women to 65 years for both sexes.

4 These revised population estimates were published on 9 September 2004 (for mid-2001 and mid-2002) and 7 October 2004 (for mid-1992 to mid-2000), following the local authority population studies, and replace all earlier versions. All figures shown on this table are now therefore on a consistent basis.

5 Data for mid-2004 for London were revised due to the Harrow correction that was published on 20 December 2005.

Tel no. for all queries relating to population estimates:- 01329 813318.

Table 1.4	Population: age and sex

Constituent countries of the United Kingdom   Numbers (thousands)

Mid-year	All ages	Age group												Under 16	16–64/59	65/60 and over
		Under 1	1–4	5–14	15–24	25–34	35–44	45–59	60–64	65–74	75–84	85–89	90 and over			
United Kingdom																
Persons																
1976	56,216	677	3,043	9,176	8,126	7,868	6,361	9,836	3,131	5,112	2,348	390	147	13,797	32,757	9,663
1981	56,357	730	2,726	8,147	9,019	8,010	6,774	9,540	2,935	5,195	2,677	..	..	12,543	33,780	10,035
1986	56,684	748	2,886	7,143	9,200	8,007	7,711	9,212	3,069	5,020	2,971	716	..	11,645	34,725	10,313
1991	57,439	790	3,077	7,141	8,168	8,898	7,918	9,500	2,888	5,067	3,119	626	248	11,685	35,197	10,557
1996[1]	58,164	719	3,019	7,544	7,231	9,131	7,958	10,553	2,785	5,066	3,129	711	317	12,018	35,498	10,649
1998[1]	58,475	713	2,930	7,649	7,079	8,948	8,285	10,767	2,835	4,979	3,211	736	344	12,013	35,746	10,717
1999[1]	58,684	704	2,896	7,684	7,090	8,795	8,474	10,887	2,877	4,948	3,230	746	354	12,011	35,928	10,745
2000[1]	58,886	682	2,869	7,652	7,139	8,646	8,678	11,011	2,900	4,940	3,249	755	364	11,959	36,138	10,788
2001[1]	59,113	663	2,819	7,624	7,261	8,475	8,846	11,168	2,884	4,947	3,296	753	377	11,863	36,406	10,845
2002[1]	59,322	661	2,753	7,601	7,403	8,256	9,002	11,316	2,890	4,969	3,345	739	388	11,783	36,622	10,916
2003	59,554	679	2,703	7,542	7,575	8,070	9,108	11,424	2,943	5,005	3,401	706	399	11,712	36,828	11,014
2004[2]	59,834	705	2,684	7,477	7,720	7,937	9,192	11,517	3,021	5,033	3,435	703	409	11,646	37,064	11,125
Males																
1976	27,360	348	1,564	4,711	4,145	3,981	3,214	4,820	1,466	2,204	775	101	31	7,083	17,167	3,111
1981	27,412	374	1,400	4,184	4,596	4,035	3,409	4,711	1,376	2,264	922	..	..	6,439	17,646	3,327
1986	27,542	384	1,478	3,664	4,663	4,022	3,864	4,572	1,463	2,206	1,060	166	..	5,968	18,142	3,432
1991	27,909	403	1,572	3,655	4,146	4,432	3,949	4,732	1,390	2,272	1,146	166	46	5,976	18,303	3,630
1996[1]	28,287	369	1,547	3,857	3,652	4,540	3,954	5,244	1,360	2,311	1,187	201	65	6,148	18,375	3,764
1998[1]	28,458	365	1,503	3,916	3,570	4,444	4,109	5,342	1,388	2,293	1,240	215	73	6,151	18,486	3,821
1999[1]	28,578	361	1,485	3,934	3,577	4,367	4,200	5,400	1,409	2,289	1,259	221	77	6,152	18,582	3,845
2000[1]	28,690	350	1,469	3,920	3,606	4,292	4,298	5,457	1,420	2,294	1,278	225	81	6,128	18,685	3,878
2001[1]	28,832	338	1,445	3,906	3,672	4,215	4,382	5,534	1,412	2,308	1,308	227	85	6,077	18,827	3,928
2002[1]	28,963	339	1,409	3,895	3,754	4,107	4,460	5,604	1,414	2,327	1,339	226	89	6,037	18,945	3,982
2003	29,108	349	1,384	3,864	3,850	4,018	4,514	5,653	1,439	2,354	1,371	219	94	6,002	19,068	4,038
2004	29,271	361	1,375	3,833	3,933	3,954	4,553	5,694	1,476	2,374	1,394	224	99	5,970	19,210	4,091
Females																
1976	28,856	330	1,479	4,465	3,980	3,887	3,147	5,015	1,665	2,908	1,573	289	116	6,714	15,590	6,552
1981	28,946	356	1,327	3,963	4,423	3,975	3,365	4,829	1,559	2,931	1,756	..	..	6,104	16,134	6,708
1986	29,142	364	1,408	3,480	4,538	3,985	3,847	4,639	1,606	2,814	1,911	550	..	5,678	16,583	6,881
1991	29,530	387	1,505	3,487	4,021	4,466	3,968	4,769	1,498	2,795	1,972	460	202	5,709	16,894	6,927
1996[1]	29,877	350	1,472	3,687	3,579	4,591	4,005	5,309	1,426	2,755	1,942	509	252	5,870	17,123	6,885
1998[1]	30,017	348	1,427	3,733	3,509	4,504	4,176	5,425	1,447	2,686	1,971	521	271	5,861	17,260	6,895
1999[1]	30,106	343	1,412	3,750	3,513	4,428	4,273	5,487	1,468	2,659	1,971	525	277	5,859	17,346	6,900
2000[1]	30,196	333	1,399	3,732	3,533	4,353	4,380	5,554	1,481	2,646	1,971	530	283	5,832	17,453	6,911
2001[1]	30,281	324	1,375	3,718	3,589	4,260	4,465	5,634	1,473	2,640	1,987	526	292	5,786	17,579	6,917
2002[1]	30,359	323	1,344	3,706	3,649	4,149	4,542	5,712	1,476	2,641	2,006	512	299	5,747	17,677	6,934
2003	30,446	331	1,319	3,677	3,725	4,052	4,594	5,771	1,504	2,651	2,030	486	305	5,710	17,760	6,976
2004[2]	30,563	343	1,309	3,644	3,787	3,983	4,640	5,823	1,545	2,659	2,041	478	310	5,676	17,854	7,034
England and Wales																
Persons																
1976	49,459	585	2,642	7,967	7,077	6,979	5,608	8,707	2,777	4,540	2,093	351	135	11,973	28,894	8,593
1981	49,634	634	2,372	7,085	7,873	7,086	5,996	8,433	2,607	4,619	2,388	383	157	10,910	29,796	8,928
1986	49,999	654	2,522	6,226	8,061	7,052	6,856	8,136	2,725	4,470	2,655	461	182	10,161	30,647	9,190
1991	50,748	698	2,713	6,248	7,165	7,862	7,022	8,407	2,553	4,506	2,790	561	223	10,247	31,100	9,400
1996[1]	51,410	637	2,668	6,636	6,336	8,076	7,017	9,363	2,457	4,496	2,801	639	285	10,584	31,353	9,474
1998[1]	51,720	631	2,594	6,740	6,212	7,925	7,304	9,552	2,503	4,411	2,875	661	311	10,599	31,591	9,530
1999[1]	51,933	625	2,566	6,779	6,228	7,800	7,475	9,656	2,542	4,381	2,891	671	319	10,608	31,771	9,554
2000[1]	52,140	607	2,544	6,757	6,275	7,682	7,661	9,764	2,564	4,372	2,907	680	328	10,572	31,977	9,591
2001[1]	52,360	589	2,502	6,740	6,387	7,536	7,816	9,898	2,549	4,377	2,947	677	340	10,495	32,226	9,639
2002[1]	52,570	589	2,445	6,726	6,520	7,349	7,962	10,027	2,553	4,395	2,990	664	351	10,435	32,435	9,700
2003	52,794	606	2,402	6,677	6,681	7,190	8,062	10,116	2,599	4,427	3,039	634	360	10,381	32,627	9,786
2004[2]	53,046	629	2,388	6,621	6,817	7,073	8,140	10,188	2,669	4,451	3,067	633	370	10,327	32,837	9,882
Males																
1976	24,089	300	1,358	4,091	3,610	3,532	2,843	4,280	1,304	1,963	690	91	29	6,148	15,169	2,773
1981	24,160	324	1,218	3,639	4,011	3,569	3,024	4,178	1,227	2,020	825	94	32	5,601	15,589	2,970
1986	24,311	335	1,292	3,194	4,083	3,542	3,438	4,053	1,302	1,972	951	115	35	5,208	16,031	3,072
1991	24,681	356	1,385	3,198	3,638	3,920	3,504	4,199	1,234	2,027	1,029	150	42	5,240	16,193	3,248
1996[1]	25,030	327	1,368	3,393	3,202	4,020	3,489	4,659	1,205	2,059	1,067	182	59	5,416	16,247	3,367
1998[1]	25,201	323	1,331	3,451	3,135	3,942	3,627	4,744	1,230	2,041	1,115	194	66	5,428	16,355	3,417
1999[1]	25,323	321	1,315	3,471	3,144	3,880	3,711	4,793	1,250	2,036	1,132	200	70	5,434	16,452	3,437
2000[1]	25,438	311	1,303	3,462	3,172	3,823	3,802	4,842	1,259	2,040	1,148	204	73	5,416	16,556	3,466
2001[1]	25,574	301	1,281	3,453	3,231	3,758	3,881	4,907	1,252	2,052	1,175	206	77	5,376	16,688	3,510
2002[1]	25,702	302	1,251	3,446	3,307	3,664	3,955	4,967	1,253	2,069	1,203	205	81	5,346	16,799	3,557
2003	25,841	311	1,230	3,422	3,394	3,588	4,006	5,008	1,274	2,092	1,231	199	85	5,320	16,914	3,607
2004	25,988	322	1,223	3,395	3,473	3,531	4,043	5,040	1,307	2,109	1,251	203	90	5,294	17,041	3,653
Females																
1976	25,370	285	1,284	3,876	3,467	3,447	2,765	4,428	1,473	2,577	1,403	261	106	5,826	13,725	5,820
1981	25,474	310	1,154	3,446	3,863	3,517	2,972	4,255	1,380	2,599	1,564	289	126	5,309	14,207	5,958
1986	25,687	319	1,231	3,032	3,978	3,509	3,418	4,083	1,422	2,498	1,704	346	148	4,953	14,616	6,118
1991	26,067	342	1,328	3,050	3,527	3,943	3,517	4,208	1,319	2,479	1,761	411	181	5,007	14,908	6,152
1996[1]	26,381	310	1,300	3,243	3,134	4,056	3,528	4,704	1,252	2,437	1,734	457	227	5,168	15,106	6,107
1998[1]	26,519	308	1,264	3,289	3,077	3,983	3,677	4,808	1,272	2,370	1,760	467	244	5,171	15,235	6,113
1999[1]	26,610	305	1,251	3,308	3,083	3,920	3,763	4,863	1,292	2,345	1,759	472	249	5,175	15,318	6,117
2000[1]	26,702	296	1,241	3,296	3,103	3,859	3,859	4,923	1,304	2,332	1,758	476	255	5,155	15,421	6,126
2001[1]	26,786	288	1,220	3,287	3,156	3,778	3,935	4,992	1,297	2,326	1,771	471	263	5,119	15,538	6,129
2002[1]	26,868	287	1,194	3,280	3,214	3,684	4,007	5,059	1,300	2,326	1,787	460	270	5,090	15,635	6,143
2003	26,953	295	1,172	3,256	3,287	3,602	4,056	5,108	1,325	2,335	1,808	436	275	5,061	15,714	6,179
2004[2]	27,057	307	1,164	3,226	3,344	3,542	4,098	5,148	1,362	2,341	1,816	429	280	5,033	15,796	6,229

Note: Figures may not add exactly due to rounding.
1   These revised population estimates were published on 9 September 2004 (for mid-2001 and mid-2002) and 7 October 2004 (for mid-1992 to mid-2000), following the local authority population studies, and replace all earlier versions. All figures shown on this table are now therefore on a consistent basis.
2   Data for mid-2004 for the United Kingdom, Great Britain, England and Wales and England were revised due to the Harrow correction that was published on 20 December 2005.

Tel no. for all enquiries relating to population estimates:- 01329 813318

Table 1.4 continued — Population: age and sex

Constituent countries of the United Kingdom — Numbers (thousands)

Mid-year	All ages	Under 1	1–4	5–14	15–24	25–34	35–44	45–59	60–64	65–74	75–84	85–89	90 and over	Under 16	16–64/59	65/60 and over
England																
Persons																
1976	46,660	551	2,491	7,513	6,688	6,599	5,298	8,199	2,616	4,274	1,972	332	127	11,293	27,275	8,092
1981	46,821	598	2,235	6,678	7,440	6,703	5,663	7,948	2,449	4,347	2,249	362	149	10,285	28,133	8,403
1986	47,188	618	2,380	5,869	7,623	6,682	6,478	7,672	2,559	4,199	2,501	435	172	9,583	28,962	8,643
1991	47,875	660	2,560	5,885	6,772	7,460	6,633	7,920	2,399	4,222	2,626	529	210	9,658	29,390	8,827
1996[1]	48,519	603	2,523	6,255	5,985	7,667	6,638	8,822	2,310	4,217	2,631	602	269	9,985	29,639	8,895
1998[1]	48,821	598	2,453	6,356	5,869	7,524	6,915	8,999	2,353	4,140	2,698	623	293	10,003	29,868	8,950
1999[1]	49,033	592	2,427	6,394	5,881	7,412	7,079	9,097	2,391	4,114	2,713	632	301	10,014	30,044	8,975
2000[1]	49,233	575	2,406	6,375	5,923	7,304	7,257	9,199	2,411	4,107	2,727	641	309	9,980	30,243	9,010
2001[1]	49,450	558	2,366	6,359	6,032	7,171	7,407	9,327	2,395	4,113	2,764	638	321	9,908	30,487	9,055
2002[1]	49,647	558	2,312	6,345	6,155	6,993	7,548	9,448	2,397	4,130	2,804	625	331	9,853	30,683	9,111
2003	49,856	575	2,273	6,300	6,304	6,843	7,643	9,533	2,438	4,159	2,852	596	340	9,804	30,862	9,190
2004[2]	50,093	597	2,260	6,247	6,432	6,732	7,718	9,600	2,503	4,181	2,879	594	349	9,754	31,059	9,280
Males																
1976	22,728	283	1,280	3,858	3,413	3,339	2,686	4,031	1,228	1,849	649	85	27	5,798	14,320	2,610
1981	22,795	306	1,147	3,430	3,790	3,377	2,856	3,938	1,154	1,902	777	89	30	5,280	14,717	2,798
1986	22,949	317	1,219	3,010	3,862	3,357	3,249	3,822	1,224	1,853	897	108	33	4,911	15,147	2,891
1991	23,291	336	1,307	3,011	3,439	3,721	3,311	3,957	1,159	1,900	970	141	39	4,938	15,302	3,050
1996[1]	23,629	309	1,294	3,198	3,023	3,818	3,302	4,390	1,133	1,932	1,003	172	55	5,110	15,358	3,161
1998[1]	23,794	306	1,258	3,254	2,960	3,743	3,436	4,470	1,157	1,916	1,047	183	62	5,123	15,462	3,209
1999[1]	23,916	304	1,243	3,274	2,969	3,689	3,517	4,516	1,176	1,913	1,063	188	66	5,129	15,558	3,229
2000[1]	24,030	294	1,232	3,266	2,995	3,638	3,604	4,562	1,184	1,917	1,078	192	69	5,113	15,661	3,256
2001[1]	24,166	285	1,212	3,257	3,053	3,580	3,681	4,624	1,176	1,928	1,103	194	73	5,075	15,793	3,298
2002[1]	24,288	286	1,183	3,251	3,123	3,492	3,753	4,682	1,176	1,944	1,128	193	77	5,047	15,899	3,342
2003	24,415	295	1,164	3,228	3,204	3,418	3,802	4,721	1,195	1,965	1,156	187	80	5,024	16,003	3,388
2004	24,554	306	1,158	3,203	3,278	3,364	3,837	4,752	1,225	1,981	1,175	191	85	5,000	16,122	3,431
Females																
1976	23,932	269	1,211	3,656	3,275	3,260	2,612	4,168	1,387	2,425	1,323	246	100	5,495	14,968	5,481
1981	24,026	292	1,088	3,248	3,650	3,327	2,807	4,009	1,295	2,445	1,472	273	119	5,004	13,416	5,605
1986	24,239	301	1,161	2,859	3,761	3,325	3,229	3,850	1,335	2,346	1,604	326	140	4,672	13,815	5,752
1991	24,584	324	1,253	2,873	3,333	3,739	3,322	3,964	1,239	2,323	1,656	388	171	4,720	14,088	5,777
1996[1]	24,890	293	1,229	3,056	2,961	3,849	3,336	4,432	1,177	2,286	1,628	430	214	4,876	14,281	5,734
1998[1]	25,027	292	1,195	3,102	2,908	3,781	3,479	4,529	1,196	2,224	1,651	440	230	4,880	14,406	5,741
1999[1]	25,117	288	1,183	3,121	2,912	3,724	3,562	4,581	1,215	2,201	1,650	444	235	4,885	14,486	5,746
2000[1]	25,203	281	1,174	3,109	2,928	3,667	3,653	4,637	1,227	2,190	1,649	448	240	4,867	14,582	5,755
2001[1]	25,284	273	1,154	3,102	2,979	3,591	3,726	4,702	1,219	2,185	1,661	444	248	4,834	14,694	5,757
2002[1]	25,358	272	1,129	3,095	3,031	3,501	3,795	4,766	1,220	2,186	1,676	433	254	4,806	14,783	5,769
2003	25,441	280	1,109	3,072	3,100	3,424	3,841	4,812	1,243	2,194	1,696	409	260	4,780	14,859	5,802
2004[2]	25,539	291	1,103	3,044	3,155	3,368	3,881	4,849	1,278	2,200	1,704	403	264	4,754	14,936	5,849
Wales																
Persons																
1976	2,799	33	151	453	388	379	309	509	161	267	121	19	7	680	1,618	501
1981	2,813	36	136	407	434	383	333	485	158	272	139	21	8	626	1,663	525
1986	2,811	37	143	357	438	369	378	464	166	271	154	26	10	578	1,686	547
1991	2,873	38	153	363	393	402	389	486	154	284	164	32	13	589	1,711	573
1996[1]	2,891	34	146	381	352	409	379	541	147	279	170	37	17	598	1,714	578
1998[1]	2,900	34	141	384	343	401	390	553	150	271	177	38	18	596	1,723	581
1999[1]	2,901	33	139	385	347	388	395	559	151	267	178	39	18	594	1,727	580
2000[1]	2,907	32	138	383	352	378	403	565	152	265	180	39	19	591	1,734	581
2001[1]	2,910	32	136	382	356	365	409	572	154	264	183	39	20	587	1,739	584
2002[1]	2,923	30	132	380	366	356	415	579	156	265	185	39	20	582	1,752	589
2003	2,938	31	129	377	377	347	418	583	161	268	187	38	20	577	1,765	596
2004	2,952	32	127	374	385	341	422	588	166	270	188	39	21	572	1,778	602
Males																
1976	1,361	17	78	233	197	193	157	249	75	114	41	5	2	350	849	162
1981	1,365	18	70	209	221	193	168	240	73	118	48	5	2	321	871	173
1986	1,362	19	73	184	221	186	190	231	79	119	54	7	2	297	885	181
1991	1,391	20	78	186	199	199	194	242	74	128	60	8	2	302	891	198
1996[1]	1,401	17	74	195	179	203	187	269	72	128	64	10	3	306	890	206
1998[1]	1,407	17	72	197	174	199	192	274	73	125	68	11	4	305	894	208
1999[1]	1,408	17	72	198	176	192	194	277	74	124	69	11	4	305	895	208
2000[1]	1,408	16	71	196	177	185	198	280	75	124	71	12	4	303	895	210
2001[1]	1,409	16	69	196	179	178	200	283	75	124	73	12	4	301	895	212
2002[1]	1,414	16	68	195	183	172	202	286	77	125	74	12	5	299	900	215
2003	1,426	16	66	194	191	170	204	287	79	127	75	12	5	297	911	219
2004	1,434	16	66	192	196	167	206	289	82	128	76	12	5	294	918	222
Females																
1976	1,438	16	73	220	191	187	153	260	86	152	80	14	6	330	770	339
1981	1,448	18	66	199	213	190	165	246	85	154	91	16	6	305	791	352
1986	1,449	18	70	173	217	184	188	233	87	152	100	20	8	282	801	366
1991	1,482	19	75	177	194	203	195	244	80	156	104	24	10	288	820	375
1996[1]	1,490	16	71	186	173	206	192	272	75	151	106	27	13	293	825	373
1998[1]	1,492	16	69	187	169	202	198	278	76	146	109	27	14	290	829	373
1999[1]	1,493	16	68	187	171	196	201	282	77	144	109	27	15	289	832	371
2000[1]	1,499	15	67	186	175	192	206	285	77	142	109	28	15	288	840	371
2001[1]	1,502	15	66	186	177	187	209	289	78	141	110	27	15	286	844	372
2002[1]	1,509	15	65	185	182	183	212	293	80	140	111	27	16	283	852	374
2003	1,512	15	63	184	186	178	214	296	82	141	112	26	16	281	855	377
2004	1,518	15	62	182	189	174	216	299	85	142	112	26	16	278	859	380

Table 1.4 continued	Population: age and sex

Constituent countries of the United Kingdom
Numbers (thousands)

Mid-year	All ages	Under 1	1–4	5–14	15–24	25–34	35–44	45–59	60–64	65–74	75–84	85–89	90 and over	Under 16	16–64/59	65/60 and over
Scotland																
Persons																
1976	5,233	67	291	904	806	692	591	897	282	460	202	31	11	1,352	3,023	858
1981	5,180	69	249	780	875	724	603	880	260	460	232	35	14	1,188	3,110	882
1986	5,112	66	257	656	863	739	665	849	273	435	252	42	15	1,061	3,161	890
1991	5,083	66	258	634	746	795	696	853	265	441	259	51	19	1,021	3,151	912
1996	5,092	59	252	643	651	798	722	925	259	448	256	57	24	1,019	3,151	922
1998	5,077	58	239	644	628	766	749	941	261	445	262	59	26	1,003	3,145	929
1999	5,072	56	234	643	625	743	762	951	262	444	265	59	27	995	3,144	933
2000	5,063	53	230	636	628	717	774	962	263	445	267	59	28	985	3,141	937
2001	5,064	52	224	629	633	696	782	979	262	447	272	59	29	970	3,150	944
2002	5,055	51	217	622	639	669	788	993	262	449	276	58	30	955	3,150	950
2003	5,057	52	212	614	648	648	793	1,008	265	452	281	55	31	943	3,156	958
2004	5,078	54	210	609	653	635	796	1,025	270	455	286	54	31	935	3,175	968
2005	5,059	54	211	600	659	629	794	1,042	273	457	286	59	32	929	3,191	975
Males																
1976	2,517	34	149	463	408	347	290	429	128	193	65	8	2	693	1,556	269
1981	2,495	35	128	400	445	364	298	424	118	194	77	8	3	610	1,603	282
1986	2,462	34	131	336	438	371	331	410	127	184	86	10	3	543	1,636	283
1991	2,445	34	132	324	377	394	345	415	124	192	91	13	3	522	1,623	299
1996	2,447	30	128	328	327	392	355	454	122	198	93	15	5	521	1,616	310
1998	2,439	30	122	329	315	374	367	463	124	198	96	16	5	513	1,610	316
1999	2,437	29	120	329	313	362	372	469	125	198	98	16	6	510	1,609	318
2000	2,432	28	118	326	315	347	377	474	125	199	100	17	6	505	1,606	322
2001	2,434	26	115	322	319	337	379	483	125	200	103	17	6	497	1,610	327
2002	2,432	26	111	319	324	325	382	490	125	202	106	17	7	489	1,612	331
2003	2,435	26	108	314	329	315	383	496	126	204	108	16	7	483	1,616	336
2004	2,446	28	107	312	332	310	384	503	129	207	111	16	7	479	1,627	341
2005	2,456	28	107	307	335	309	382	511	131	208	112	18	7	476	1,635	345
Females																
1976	2,716	32	142	440	398	345	301	468	154	267	137	23	8	659	1,468	589
1981	2,685	33	121	380	430	359	305	456	142	265	155	27	11	579	1,506	600
1986	2,649	32	126	320	424	368	334	439	146	250	166	32	12	518	1,525	606
1991	2,639	32	126	309	369	402	351	437	141	249	168	38	16	499	1,528	612
1996	2,645	28	123	315	324	406	367	470	137	250	164	42	20	498	1,535	612
1998	2,638	28	116	315	313	392	382	478	137	248	166	43	21	490	1,535	614
1999	2,635	27	114	314	312	381	390	483	138	246	166	43	22	486	1,535	614
2000	2,631	26	112	310	313	369	397	488	138	246	166	43	22	480	1,535	616
2001	2,630	26	109	307	314	359	403	496	137	246	169	43	23	473	1,540	617
2002	2,623	25	106	303	315	344	406	504	137	247	171	41	23	466	1,538	619
2003	2,623	25	104	300	318	332	410	512	139	248	173	39	24	460	1,540	622
2004	2,632	26	103	297	321	325	412	521	141	248	175	38	24	457	1,549	627
2005	2,639	26	103	293	324	320	411	531	142	249	174	41	25	453	1,556	630
Northern Ireland																
Persons																
1976	1,524	26	111	306	243	198	163	231	73	111	53	8	2	471	840	212
1981	1,543	27	106	282	271	200	175	227	68	116	57	..	..	444	874	224
1986	1,574	28	107	261	277	217	190	227	71	115	64	16	..	423	917	234
1991	1,607	26	106	260	256	240	200	241	70	121	69	14	6	417	945	246
1996	1,662	24	99	266	244	257	220	266	70	123	72	15	7	415	993	253
1998	1,678	24	97	264	239	257	231	275	71	122	74	16	7	411	1,010	257
1999	1,679	23	96	262	237	252	237	279	73	122	75	16	7	408	1,014	258
2000	1,683	22	95	259	237	247	243	284	73	123	75	16	7	403	1,020	259
2001	1,689	22	93	255	240	243	248	290	74	123	77	16	7	397	1,030	262
2002	1,697	22	91	253	243	238	251	296	75	125	79	16	7	393	1,037	266
2003	1,703	21	89	251	246	233	254	301	78	126	81	16	8	388	1,044	271
2004	1,710	22	87	248	250	229	256	305	81	127	82	16	8	383	1,052	275
Males																
1976	754	13	58	157	127	102	81	111	34	47	19	3	-	242	442	70
1981	757	14	54	145	140	102	87	109	32	50	21	..	..	228	454	75
1986	768	14	55	134	142	109	95	110	33	50	23	4	..	217	474	77
1991	783	13	54	133	131	119	100	118	32	53	26	4	1	213	487	83
1996	810	12	51	136	124	128	109	131	33	54	27	4	1	212	511	87
1998	819	12	50	135	121	128	114	135	34	54	28	5	2	211	520	89
1999	818	12	49	134	119	125	117	138	35	54	29	5	2	209	521	89
2000	820	11	49	133	120	122	119	141	35	55	29	5	2	207	524	90
2001	824	11	48	131	122	120	122	144	35	56	30	5	2	204	529	92
2002	829	11	47	130	124	117	123	147	36	56	31	5	2	202	534	94
2003	833	11	46	129	126	115	124	149	38	58	31	5	2	199	538	95
2004	836	11	45	127	128	113	125	151	39	58	32	5	2	197	542	97
Females																
1976	769	13	53	149	116	96	81	120	38	64	33	6	2	229	398	143
1981	786	13	52	137	130	98	88	118	37	66	37	..	..	216	420	150
1986	805	13	52	127	135	107	96	118	38	65	41	12	..	206	442	157
1991	824	13	52	127	125	121	100	123	38	67	44	10	4	203	458	163
1996	851	11	49	130	120	129	110	135	37	69	45	11	6	203	482	167
1998	859	12	47	129	118	129	117	139	37	68	46	11	6	201	490	168
1999	861	11	47	128	117	127	120	141	38	68	46	11	6	199	493	169
2000	862	11	46	126	118	125	124	143	38	68	46	11	6	196	497	169
2001	865	10	45	124	119	123	126	146	38	68	47	11	6	193	501	170
2002	868	11	44	123	119	120	128	149	39	68	48	11	6	191	504	173
2003	870	10	43	122	120	118	129	152	40	68	49	11	6	189	506	175
2004	874	11	42	121	122	116	130	154	42	69	50	11	6	187	509	178

Table 1.5 — Population: age, sex and legal marital status[1]

England and Wales — Numbers (thousands)

Mid-year	Total population	Males					Females				
		Single	Married	Divorced	Widowed	Total	Single	Married	Divorced	Widowed	Total
Aged											
16 and over											
1971	36,818	4,173	12,522	187	682	17,563	3,583	12,566	296	2,810	19,255
1976	37,486	4,369	12,511	376	686	17,941	3,597	12,538	533	2,877	19,545
1981	38,724	5,013	12,238	611	698	18,559	4,114	12,284	828	2,939	20,165
1986	39,837	5,625	11,867	917	695	19,103	4,617	12,000	1,165	2,953	20,734
1991	40,501	5,891	11,636	1,187	727	19,441	4,817	11,833	1,459	2,951	21,060
1996	40,827	6,225	11,310	1,346	733	19,614	5,168	11,433	1,730	2,881	21,212
1997	40,966	6,337	11,240	1,379	734	19,690	5,288	11,353	1,781	2,855	21,276
1998	41,121	6,450	11,183	1,405	735	19,773	5,406	11,284	1,827	2,832	21,349
1999	41,325	6,582	11,143	1,433	732	19,890	5,526	11,235	1,875	2,800	21,435
2000	41,569	6,721	11,113	1,456	731	20,022	5,650	11,199	1,927	2,772	21,547
2001	41,865	6,894	11,090	1,482	733	20,198	5,798	11,150	1,975	2,745	21,667
2002	42,135	7,076	11,015	1,535	731	20,357	5,961	11,073	2,035	2,709	21,778
2003	42,413	7,261	10,940	1,590	728	20,520	6,128	11,000	2,096	2,668	21,892
2004	42,719	7,461	10,863	1,644	726	20,694	6,306	10,935	2,156	2,628	22,025
16–19											
1971	2,666	1,327	34	0	0	1,362	1,163	142	0	0	1,305
1976	2,901	1,454	28	0	0	1,482	1,289	129	0	0	1,419
1981	3,310	1,675	20	0	0	1,694	1,523	93	0	0	1,616
1986	3,131	1,587	10	0	0	1,596	1,484	49	1	0	1,535
1991	2,665	1,358	8	0	0	1,366	1,267	32	0	0	1,300
1996	2,402	1,209	6	0	0	1,216	1,164	21	0	0	1,186
1997	2,478	1,246	6	0	0	1,253	1,203	20	1	1	1,225
1998	2,532	1,274	6	1	0	1,281	1,230	20	1	1	1,251
1999	2,543	1,280	6	1	1	1,288	1,234	20	1	1	1,255
2000	2,523	1,276	6	1	1	1,283	1,221	18	1	1	1,240
2001	2,567	1,304	5	1	1	1,312	1,237	16	1	1	1,255
2002	2,633	1,347	4	1	1	1,353	1,266	13	1	1	1,280
2003	2,702	1,386	4	1	1	1,391	1,299	12	0	1	1,311
2004	2,770	1,423	3	0	0	1,427	1,332	11	0	0	1,343
20–24											
1971	3,773	1,211	689	3	0	1,904	745	1,113	9	2	1,869
1976	3,395	1,167	557	4	0	1,728	725	925	16	2	1,667
1981	3,744	1,420	466	10	1	1,896	1,007	811	27	2	1,847
1986	4,171	1,768	317	14	0	2,099	1,383	657	32	1	2,072
1991	3,911	1,717	242	12	0	1,971	1,421	490	29	1	1,941
1996	3,291	1,538	117	3	0	1,658	1,361	260	11	1	1,633
1997	3,141	1,479	99	3	0	1,580	1,325	225	9	1	1,561
1998	3,047	1,442	86	2	0	1,530	1,306	201	8	1	1,517
1999	3,047	1,449	78	2	0	1,530	1,320	188	8	1	1,517
2000	3,088	1,470	74	3	0	1,548	1,352	180	8	1	1,540
2001	3,157	1,501	74	3	1	1,579	1,390	178	8	1	1,578
2002	3,211	1,534	69	3	1	1,607	1,428	166	8	1	1,604
2003	3,283	1,573	69	3	1	1,646	1,466	161	8	1	1,637
2004	3,358	1,621	67	3	1	1,692	1,499	156	8	2	1,665
25–29											
1971	3,267	431	1,206	16	1	1,654	215	1,367	29	4	1,614
1976	3,758	533	1,326	39	2	1,900	267	1,522	65	5	1,859
1981	3,372	588	1,057	54	1	1,700	331	1,247	89	4	1,671
1986	3,713	835	949	79	1	1,863	527	1,207	113	4	1,850
1991	4,154	1,132	856	82	1	2,071	800	1,158	123	2	2,083
1996	3,950	1,273	650	46	1	1,970	977	906	93	3	1,980
1997	3,877	1,294	595	42	1	1,932	1,012	844	85	3	1,945
1998	3,789	1,304	544	38	1	1,887	1,039	783	77	3	1,902
1999	3,687	1,304	497	34	1	1,836	1,051	725	72	3	1,851
2000	3,605	1,305	459	31	1	1,796	1,065	677	65	3	1,810
2001	3,487	1,293	420	28	1	1,742	1,059	625	58	3	1,745
2002	3,348	1,276	371	26	1	1,674	1,052	567	52	3	1,674
2003	3,262	1,271	337	25	1	1,634	1,053	524	49	2	1,628
2004	3,260	1,292	318	24	1	1,635	1,080	497	47	2	1,625

| Table 1.5 continued | Population: age, sex and legal marital status[1] |

England and Wales
Numbers (thousands)

Mid-year	Total population	Males					Females				
		Single	Married	Divorced	Widowed	Total	Single	Married	Divorced	Widowed	Total
30–34											
1971	2,897	206	1,244	23	3	1,475	111	1,269	34	8	1,422
1976	3,220	236	1,338	55	3	1,632	118	1,388	75	8	1,588
1981	3,715	318	1,451	97	3	1,869	165	1,544	129	9	1,846
1986	3,338	355	1,197	124	2	1,679	206	1,293	154	6	1,660
1991	3,708	520	1,172	155	2	1,849	335	1,330	189	5	1,859
1996	4,126	776	1,135	138	2	2,050	551	1,316	201	7	2,076
1997	4,151	817	1,111	133	2	2,064	589	1,293	198	7	2,088
1998	4,136	848	1,078	127	3	2,056	621	1,259	193	7	2,081
1999	4,113	877	1,043	121	3	2,044	651	1,223	188	7	2,069
2000	4,076	904	1,007	114	2	2,027	679	1,182	181	7	2,049
2001	4,050	934	971	108	2	2,016	711	1,142	174	7	2,033
2002	4,000	961	921	105	2	1,990	743	1,094	167	6	2,010
2003	3,928	981	868	102	2	1,954	767	1,043	159	6	1,974
2004	3,813	987	811	97	2	1,897	777	985	149	5	1,916
35–44											
1971	5,736	317	2,513	48	13	2,891	201	2,529	66	48	2,845
1976	5,608	286	2,442	104	12	2,843	167	2,427	129	42	2,765
1981	5,996	316	2,519	178	12	3,024	170	2,540	222	41	2,972
1986	6,856	396	2,738	293	12	3,438	213	2,815	350	39	3,418
1991	7,022	477	2,632	384	11	3,504	280	2,760	444	34	3,517
1996	7,017	653	2,426	398	12	3,489	427	2,568	497	36	3,528
1997	7,155	708	2,433	403	12	3,556	472	2,580	511	36	3,599
1998	7,304	768	2,442	405	13	3,627	522	2,596	523	36	3,677
1999	7,475	832	2,459	408	13	3,711	577	2,617	533	37	3,763
2000	7,661	899	2,481	410	12	3,802	635	2,640	547	37	3,859
2001	7,816	963	2,494	411	12	3,881	692	2,649	558	36	3,935
2002	7,962	1,031	2,489	424	12	3,955	751	2,650	571	35	4,007
2003	8,062	1,089	2,471	435	12	4,006	805	2,634	583	34	4,056
2004	8,140	1,142	2,445	444	11	4,043	858	2,614	593	32	4,098
45–64											
1971	11,887	502	4,995	81	173	5,751	569	4,709	125	733	6,136
1976	11,484	496	4,787	141	160	5,583	462	4,568	188	683	5,901
1981	11,040	480	4,560	218	147	5,405	386	4,358	271	620	5,635
1986	10,860	461	4,422	331	141	5,355	327	4,220	388	570	5,505
1991	10,960	456	4,394	456	127	5,433	292	4,211	521	503	5,527
1996	11,820	528	4,587	628	121	5,864	318	4,466	732	440	5,956
1997	11,927	545	4,593	656	120	5,914	328	4,486	770	430	6,014
1998	12,055	565	4,608	681	121	5,974	340	4,512	807	422	6,080
1999	12,198	589	4,627	706	121	6,043	355	4,541	844	415	6,155
2000	12,328	615	4,638	727	121	6,101	372	4,564	881	410	6,227
2001	12,447	644	4,647	747	121	6,159	391	4,578	918	401	6,289
2002	12,580	671	4,649	780	120	6,220	413	4,596	960	391	6,359
2003	12,715	702	4,647	815	118	6,283	437	4,613	1,002	380	6,433
2004	12,857	736	4,644	850	117	6,347	465	4,628	1,045	371	6,510
65 and over											
1971	6,592	179	1,840	17	492	2,527	580	1,437	32	2,016	4,065
1976	7,119	197	2,033	33	510	2,773	569	1,579	60	2,138	4,347
1981	7,548	216	2,167	54	534	2,971	533	1,692	90	2,263	4,578
1986	7,768	223	2,234	76	539	3,072	477	1,759	127	2,333	4,696
1991	8,080	231	2,332	99	586	3,248	422	1,853	152	2,405	4,832
1996	8,221	247	2,390	134	597	3,367	369	1,897	196	2,393	4,854
1997	8,237	248	2,404	143	597	3,391	358	1,904	207	2,377	4,845
1998	8,258	250	2,418	152	597	3,417	348	1,913	218	2,362	4,841
1999	8,262	251	2,431	161	594	3,437	338	1,922	230	2,336	4,825
2000	8,287	252	2,449	171	593	3,466	327	1,938	243	2,313	4,821
2001	8,342	254	2,478	183	595	3,510	318	1,960	259	2,295	4,832
2002	8,400	256	2,511	197	595	3,557	308	1,987	276	2,272	4,843
2003	8,461	258	2,544	211	594	3,607	301	2,015	294	2,244	4,854
2004	8,520	259	2,575	225	593	3,653	293	2,044	314	2,216	4,867

1 Marital Status Estimates for 1992 to 2002 were revised in light of the local authority population studies published 7 October 2004.

See 'Notes to Tables'.

Table 1.6	Components of population change

Constituent countries of the United Kingdom   Numbers (thousands)

Components of change (mid-year to mid-year or annual averages). Under "Net civilian migration": Total[1], To/from rest of UK, To/from Irish Republic, To/from rest of the world.

Mid-year to mid-year	Population at start of period	Total annual change	Live births	Deaths	Natural change (Live births – deaths)	Net civilian migration Total[1]	To/from rest of UK	To/from Irish Republic	To/from rest of the world	Other changes	Population at end of period
United Kingdom											
1971–76	55,928	+ 58	766	670	+ 96	– 55	–		– 55	+ 16	56,216
1976–81	56,216	+ 27	705	662	+ 42	– 33	–		– 33	+ 18	56,352
1981–86	56,357	+ 65	733	662	+ 70	– 5	–		..	..	56,684
1986–91	56,684	+148	782	647	+135	+ 13	–		..	..	57,439
1991–96	57,439	+145	756	639	+117	+ 29	–		..	..	58,164
1996–97[2]	58,164	+150	740	637	+103	+ 47	–		..	..	58,314
1997–98[2]	58,314	+161	718	617	+100	+ 60	–		..	..	58,475
1998–99[2]	58,475	+209	713	634	+ 77	+133	–		..	..	58,684
1999–2000[2]	58,684	+202	688	626	+ 62	+139	–		..	..	58,886
2000–01[2]	58,886	+227	674	599	+ 74	+153	–		..	..	59,113
2001–02[2]	59,113	+208	663	601	+ 62	+146	–		..	..	59,322
2002–03[2]	59,322	+232	682	605	+ 77	+155	–		..	..	59,554
2003–04	59,554	+281	707	603	+104	+177	–		..	..	59,834
England and Wales											
1971–76	49,152	+ 61	644	588	+ 76	– 28	+ 10	– 9	– 29	+ 13	49,459
1976–81	49,459	+ 35	612	582	+ 30	– 9	+ 11	– 3	– 17	+ 14	49,634
1981–86	49,634	+ 73	639	582	+ 57	+ 16	..	..	..	..	49,999
1986–91	49,999	+150	689	569	+120	+ 30	..	..	..	..	50,748
1991–96	50,748	+132	668	563	+106	+ 27	..	..	..	..	51,410
1996–97[2]	51,410	+149	655	562	+ 93	+ 56	..	..	..	..	51,560
1997–98[2]	51,560	+160	636	544	+ 92	+ 68	..	..	..	..	51,720
1998–99[2]	51,720	+213	630	558	+ 72	+141	..	..	..	..	51,933
1999–2000[2]	51,933	+207	612	550	+ 61	+146	..	..	..	..	52,140
2000–01[2]	52,140	+220	599	528	+ 71	+149	..	..	..	..	52,360
2001–02[2]	52,360	+210	591	530	+ 61	+149	..	..	..	..	52,570
2002–03[2]	52,570	+223	608	532	+ 76	+147	..	..	..	..	52,794
2003–04	52,794	+252	631	531	+101	+151	..	..	..	..	53,046
England											
1971–76	46,412	+ 50	627	552	+ 75	– 35	+ 1	– 9	– 27	+ 10	46,660
1976–81	46,660	+ 32	577	546	+ 31	– 11	+ 6	– 3	– 15	+ 12	46,821
1981–86	46,821	+ 73	603	547	+ 56	+ 18	..	..	..	..	47,188
1986–91	47,188	+137	651	535	+116	+ 21	..	..	..	..	47,875
1991–96	47,875	+129	632	528	+104	+ 24	..	..	..	..	48,519
1996–97[2]	48,519	+146	620	527	+ 93	+ 53	..	..	..	..	48,665
1997–98[2]	48,665	+156	602	510	+ 92	+ 64	..	..	..	..	48,821
1998–99[2]	48,821	+212	598	523	+ 74	+138	..	..	..	..	49,033
1999–2000[2]	48,033	+200	580	516	+ 64	+136	..	..	..	..	49,233
2000–01[2]	49,233	+216	568	495	+ 73	+144	..	..	..	..	49,450
2001–02[2]	49,450	+197	560	497	+ 63	+134	..	..	..	..	49,647
2002–03[2]	49,647	+209	578	498	+ 79	+130	..	..	..	..	49,856
2003–04	49,856	+237	600	498	+102	+136	..	..	..	..	50,093
Wales											
1971–76	2,740	+ 12	37	36	+ 1	+ 7	+ 10		– 2	+ 3	2,799
1976–81	2,799	+ 3	35	36	– 1	+ 2	+ 5		– 2	+ 2	2,813
1981–86	2,813	– 1	36	35	+ 1	– 1	..	..	..	..	2,811
1986–91	2,811	+ 12	38	34	+ 4	+ 8	..	..	..	..	2,873
1991–96	2,873	+ 4	36	35	+ 1	+ 2	..	..	..	..	2,891
1996–97[2]	2,891	+ 4	35	35	–	+ 3	..	..	..	..	2,895
1997–98[2]	2,895	+ 5	34	34	–	+ 4	..	..	..	..	2,900
1998–99[2]	2,900	+ 1	33	35	– 2	+ 3	..	..	..	..	2,901
1999–2000[2]	2,901	+ 6	31	34	– 3	+ 9	..	..	..	..	2,907
2000–01[2]	2,907	+ 3	31	33	– 2	+ 5	..	..	..	..	2,910
2001–02[2]	2,910	+ 13	30	33	– 3	+ 16	..	..	..	..	2,923
2002–03[2]	2,923	+ 15	31	33	– 3	+ 17	..	..	..	..	2,938
2003–04	2,938	+ 14	32	33	– 1	+ 16	..	..	..	..	2,952
Scotland											
1971–76	5,236	–	73	64	+ 9	– 14	– 4		– 10	+ 4	5,233
1976–81	5,233	– 11	66	64	+ 2	– 16	– 7		– 10	+ 4	5,180
1981–86	5,180	– 14	66	64	+ 2	– 16	– 7		– 7	+ 1	5,112
1986–91	5,112	– 6	66	62	+ 3	– 9	..	..	..	..	5,083
1991–96	5,083	+ 2	63	61	+ 1	– 0	..	..	..	..	5,092
1996–97	5,092	– 9	60	60	–	– 9	..	..	..	..	5,083
1997–98	5,083	– 6	58	59	– 1	– 6	..	..	..	..	5,077
1998–99	5,077	– 5	57	60	– 4	– 1	..	..	..	..	5,072
1999–2000	5,072	– 9	54	60	– 6	– 3	..	..	..	..	5,063
2000–01	5,063	+ 1	53	57	– 4	+ 5	..	..	..	..	5,064
2001–02	5,064	– 9	51	57	– 6	– 3	..	..	..	..	5,055
2002–03	5,055	+ 3	52	58	– 7	+ 9	..	..	..	..	5,057
2003–04	5,057	+ 21	54	58	– 4	+ 25	..	..	..	..	5,078
Northern Ireland											
1971–76	1,540	– 3	28	17	+ 11	– 14	– 7	– 7		– 1	1,524
1976–81	1,524	+ 3	27	17	+ 10	– 8	– 4	– 3		+ 17	1,543
1981–86	1,543	+ 6	28	16	+ 12	– 5	– 3	– 1		–	1,574
1986–91	1,574	+ 7	27	16	+ 12	– 5	– 3	– 1		–	1,607
1991–96	1,607	+ 11	25	15	+ 9	+ 2	..	..		–	1,662
1996–97	1,662	+ 10	25	15	+ 10	– 1	..	..		+ 1	1,671
1997–98	1,671	+ 7	24	15	+ 9	– 2	..	..		–	1,678
1998–99	1,678	+ 1	23	15	+ 8	– 5	..	..		– 2	1,679
1999–2000	1,679	+ 4	22	16	+ 7	– 2	..	..		– 1	1,683
2000–01	1,683	+ 6	22	14	+ 7	– 2	..	..		+ 1	1,689
2001–02	1,689	+ 7	21	14	+ 7		..	..		–	1,697
2002–03	1,697	+ 6	21	15	+ 7	– 1	..	..		..	1,703
2003–04	1,703	+ 8	22	15	+ 7	– 0	..	..		..	1,710

1 For UK, England, Wales and Scotland from 1981 onwards, this column is not an estimate of net civilian migration; it also includes "other" changes. It has been derived by subtraction using revised population estimates and natural change.
2 These revised population estimates were published on 9 September 2004 (for mid-2001 and mid-2002) and 7 October 2004 (for mid-1992 to mid-2000), following the local authority population studies, and replace all earlier versions. All figures shown on this table are now therefore on a consistent basis.

Table 2.1 Vital statistics summary

Constituent countries of the United Kingdom

Numbers (thousands) and rates

Year and quarter	All live births		Live births outside marriage		Marriages		Divorces		Deaths		Infant mortality[5]		Neonatal mortality[6]		Perinatal mortality[7]	
	Number	Rate[1]	Number	Rate[2]	Number	Rate[3]	Number	Rate[4]	Number	Rate[1]	Number	Rate[2]	Number	Rate[2]	Number	Rate[8]
United Kingdom																
1976	675.5	12.0	61.1	90	406.0	..	135.4	..	680.8	12.1	9.79	14.5	6.68	9.9	12.25	18.0
1981	730.7	13.0	91.3	125	397.8	49.4	156.4	11.3	658.0	11.7	8.16	11.2	4.93	6.7	8.79	12.0
1986	754.8	13.3	154.3	204	393.9	..	168.2	..	660.7	11.7	7.18	9.5	4.00	5.3	7.31	9.6
1991	792.3	13.8	236.1	298	349.7	..	173.5	..	646.2	11.2	5.82	7.4	3.46	4.4	6.45	8.1
1996	733.2	12.6	260.4	355	317.5	..	171.7	..	636.0	10.9	4.50	6.1	3.00	4.1	6.41	8.7
1999	700.0	11.9	271.6	388	301.1	..	158.7	..	632.1	10.8	4.05	5.8	2.73	3.9	5.79	8.2
2000	679.0	11.5	268.1	395	305.9	..	154.6	..	608.4	10.3	3.79	5.6	2.63	3.9	5.56	8.1
2001	669.1	11.3	268.0	401	286.1	..	156.8	..	602.3	10.2	3.66	5.5	2.43	3.6	5.39	8.0
2002	668.8	11.3	271.7	406	293.0	..	160.5	..	606.2	10.2	3.50	5.2	2.36	3.5	5.57	8.3
2003	695.6	11.7	288.5	415	308.6	..	166.7	..	612.0	10.3	3.69	5.3	2.53	3.6	5.94	8.5
2004	716.0	12.0	302.6	423	311.2[P]	..	167.1[P]	..	583.1	9.7	3.61[P]	5.0	2.46[P]	3.4	5.85	8.1
2005	722.6[P]	12.0[P]	310.2[P]	429[P]	..	..	..	..	583.0[P]	9.7[P]	3.67[P]	5.1[P]	2.52[P]	3.5[P]	..	..
2003 March	165.6	11.3	68.7	415	38.2	..	42.6	..	162.5	11.1	0.96	5.8	0.65	3.9	1.45	8.7
June	173.4	11.7	70.3	405	85.9	..	42.0	..	145.8	9.8	0.88	5.0	0.60	3.4	1.49	8.5
Sept	182.2	12.2	75.7	415	127.0	..	41.3	..	140.7	9.4	0.89	4.9	0.62	3.4	1.52	8.3
Dec	174.3	11.6	73.6	423	56.1	..	40.8	..	162.2	10.8	0.96	5.5	0.66	3.8	1.49	8.5
2004 March	174.3	11.7	73.6	422	39.7[P]	..	43.1[P]	..	159.7	10.7	0.97	5.5	0.64	3.7	1.50	8.5
June	176.2	11.8	73.2	415	85.5[P]	..	41.5[P]	..	139.3	9.4	0.84	4.8	0.59	3.4	1.45	8.2
Sept	185.1	12.3	78.5	424	128.4[P]	..	42.3[P]	..	135.1	9.0	0.90	4.9	0.64	3.5	1.55	8.3
Dec	180.4	12.0	77.3	429	57.6[P]	..	40.2[P]	..	149.0	9.9	0.90	5.0	0.58	3.2	1.36	7.5
2005 March	173.2[P]	11.7[P]	74.5[P]	430[P]	..	..	..	..	164.7[P]	11.1[P]	0.91[P]	5.3[P]	0.63[P]	3.6[P]	1.32[P]	7.6[P]
June	179.0[P]	11.9[P]	75.0[P]	419[P]	..	..	..	..	143.3[P]	9.5[P]	0.94[P]	5.3[P]	0.63[P]	3.5[P]	1.41[P]	7.9[P]
Sept	190.3[P]	12.6[P]	82.5[P]	434[P]	..	..	..	..	131.6[P]	8.7[P]	0.92[P]	4.8[P]	0.66[P]	3.5[P]	1.42[P]	7.5[P]
Dec	180.1[P]	11.9[P]	78.2[P]	434[P]	..	..	..	..	143.3[P]	9.4[P]	0.90[P]	5.0[P]	0.60[P]	3.3[P]	..	..
England and Wales																
1976	584.3	11.8	53.8	92	358.6	57.7	126.7	10.1	598.5	12.1	8.34	14.3	5.66	9.7	10.45	17.7
1981	634.5	12.8	81.0	128	352.0	49.6	145.7	11.9	577.9	11.6	7.02	11.1	4.23	6.7	7.56	11.8
1986	661.0	13.2	141.3	214	347.9	43.6	153.9	12.9	581.2	11.6	6.31	9.6	3.49	5.3	6.37	9.6
1991	699.2	13.8	211.3	302	306.8	36.0	158.7	13.5	570.0	11.2	5.16	7.4	3.05	4.4	5.65	8.0
1996	649.5	12.6	232.7	358	279.0	30.9	157.1	13.8	560.1	10.9	3.99	6.1	2.68	4.1	5.62	8.6
1999	621.9	12.0	241.9	389	263.5	27.8	144.6	12.9	556.1	10.7	3.62	5.8	2.44	3.9	5.14	8.2
2000	604.4	11.6	238.6	395	268.0	27.8	141.1	12.7	535.7	10.3	3.38	5.6	2.34	3.9	4.96	8.2
2001	594.6	11.4	238.1	400	249.2	25.4	143.8	12.9	530.4	10.1	3.24	5.4	2.14	3.6	4.76	8.0
2002	596.1	11.3	242.0	406	255.6	25.6	147.7	13.4	533.5	10.1	3.13	5.2	2.13	3.6	4.99	8.3
2003	621.5	11.8	257.2	414	270.1	26.4	153.5	14.0	538.3	10.2	3.31	5.3	2.26	3.6	5.34	8.5
2004	639.7	12.1	269.7	422	270.7[P]	25.9[P]	153.4[P]	14.1	512.5	9.7	3.22	5.0	2.21	3.5	5.23	8.1
2005	645.8	12.1[P]	276.5	428	..	..	138.6[P]	12.7[P]	513.0[P]	9.6[P]	3.25[P]	5.0[P]	2.21[P]	3.4[P]	..	..
2003 March	147.4	11.3	61.0	414	34.0	13.5	39.4	14.6	143.0	11.0	0.86	5.9	0.60	3.9	1.32	8.9
June	155.1	11.8	62.8	405	75.2	29.4	38.6	14.1	128.3	9.7	0.80	5.1	0.55	3.5	1.34	8.6
Sept	162.9	12.2	67.6	415	111.9	43.4	37.9	13.7	123.9	9.3	0.79	4.8	0.55	3.4	1.36	8.3
Dec	156.0	11.7	65.8	422	49.1	19.0	37.6	13.6	143.1	10.8	0.86	5.5	0.59	3.7	1.32	8.4
2004 March	155.2	11.8	65.2	421	35.0[P]	13.4[P]	39.5[P]	14.6[P]	140.5	10.7	0.87	5.6	0.58	3.8	1.33	8.5
June	157.4	11.9	65.2	414	74.3[P]	28.6[P]	38.1[P]	14.0[P]	122.1	9.3	0.74	4.7	0.52	3.3	1.29	8.1
Sept	165.4	12.4	70.2	424	112.2[P]	42.7[P]	39.0[P]	14.2[P]	118.6	8.9	0.80	4.8	0.57	3.5	1.39	8.4
Dec	161.7	12.1	69.4	427	49.2[P]	18.7[P]	36.9[P]	13.5[P]	131.3	9.8	0.81	5.0	0.53	3.3	1.23	7.6
2005 March	154.3	11.7[P]	66.3	430	..	..	36.2[P]	13.5[P]	145.3[P]	11.0[P]	0.82[P]	5.3[P]	0.56[P]	3.6[P]	1.18[P]	7.6[P]
June	159.7	12.0[P]	66.6	417	..	..	36.5[P]	13.4[P]	125.9[P]	9.5[P]	0.83[P]	5.2[P]	0.56[P]	3.5[P]	1.24[P]	7.8[P]
Sept	170.2	12.6[P]	73.7	433	..	..	35.6[P]	13.0[P]	115.4[P]	8.6[P]	0.80[P]	4.7[P]	0.57[P]	3.4[P]	1.27[P]	7.5[P]
Dec	161.7	12.0[P]	69.9	433	..	..	30.4[P]	11.1[P]	126.2[P]	9.4[P]	0.80[P]	4.9[P]	0.53[P]	3.3[P]	..	..
England																
1976	550.4	11.8	50.8	92	339.0	..	..	..	560.3	12.0	7.83	14.2	5.32	9.7	9.81	17.6
1981	598.2	12.8	76.9	129	332.2	..	..	..	541.0	11.6	6.50	10.9	3.93	6.6	7.04	11.7
1986	623.6	13.2	133.5	214	328.4	..	146.0	..	544.5	11.6	5.92	9.5	3.27	5.2	5.98	9.5
1991	660.8	13.7	198.9	301	290.1	..	150.1	..	534.0	11.2	4.86	7.3	2.87	4.3	5.33	8.0
1996	614.2	12.7	218.2	355	264.2	..	148.7	..	524.0	10.8	3.74	6.1	2.53	4.1	5.36	8.7
1999	589.5	12.0	226.7	385	249.5	..	137.0	..	519.6	10.8	3.38	5.7	2.29	3.9	4.86	8.2
2000	572.8	11.7	223.8	391	253.8	..	133.9	..	501.0	10.2	3.18	5.6	2.21	3.9	4.69	8.2
2001	563.7	11.4	223.3	396	236.2	..	136.4	..	496.1	10.0	3.04	5.4	2.02	3.6	4.51	8.0
2002	565.7	11.4	227.0	401	242.1	..	140.2	..	499.1	10.1	2.97	5.2	2.02	3.6	4.75	8.3
2003	589.9	11.8	241.4	409	255.6	..	145.8	..	503.4	10.1	3.14	5.3	2.15	3.7	5.01	8.5
2004	607.2	12.1	253.1	417	255.9[P]	..	145.5[P]	..	479.2	9.6	3.03	5.0	2.09	3.4	4.96	8.1
2005	613.0	12.2[P]	259.4	423	..	..	131.6[P]	..	479.7[P]	9.5[P]	3.08[P]	5.0[P]	2.11[P]	3.4[P]	..	..
2003 March	139.9	11.4	57.2	409	32.3	..	37.5	..	133.8	10.9	0.83	5.9	0.55	3.9	1.25	8.9
June	147.3	11.8	58.9	400	71.2	..	36.6	..	119.6	9.1	0.76	5.1	0.52	3.6	1.28	8.6
Sept	154.5	12.3	63.4	411	105.6	..	36.0	..	116.0	8.7	0.74	4.8	0.52	3.3	1.28	8.3
Dec	148.2	11.8	61.8	417	46.5	..	35.7	..	134.0	10.1	0.82	5.5	0.56	3.8	1.26	8.4
2004 March	147.3	11.8	61.2	416	33.2[P]	..	37.4[P]	..	131.4	10.6	0.82	5.5	0.55	3.7	1.25	8.4
June	149.6	12.0	61.3	410	70.3[P]	..	36.0[P]	..	114.2	9.2	0.69	4.6	0.49	3.3	1.22	8.1
Sept	156.9	12.5	65.8	420	105.8[P]	..	36.9[P]	..	110.8	8.8	0.74	4.7	0.53	3.4	1.31	8.3
Dec	153.3	12.2	64.7	422	46.5[P]	..	35.1[P]	..	122.9	9.8	0.78	5.1	0.52	3.4	1.17	7.6
2005 March	146.4	11.8[P]	62.1	424	..	..	34.4[P]	..	135.8[P]	10.9[P]	0.78[P]	5.3[P]	0.53[P]	3.6[P]	1.17[P]	8.0[P]
June	151.8	12.1[P]	62.5	412	..	..	34.7[P]	..	117.7[P]	9.4[P]	0.79[P]	5.2[P]	0.53[P]	3.5[P]	1.18[P]	7.7[P]
Sept	161.4	12.7[P]	69.1	428	..	..	33.8[P]	..	108.0[P]	8.5[P]	0.76[P]	4.7[P]	0.54[P]	3.4[P]	1.21[P]	7.5[P]
Dec	153.4	12.1[P]	65.6	428	..	..	28.7[P]	..	118.1[P]	9.3[P]	0.75[P]	4.9[P]	0.51[P]	3.3[P]	..	..

Notes: Rates for the most recent quarters will be particularly subject to revision, even when standard detail is given, as they are based on provisional numbers or on estimates derived from events registered in the period.

Figures for England and Wales represent the number of deaths registered in each year up to 1992, and the number of deaths occurring in each year from 1993 to 2004. Provisional figures for 2005 relate to registrations. Death rates for 2005 are based on 2004-based population projections for 2005.

Birth and death figures for England and also for Wales each exclude events for persons usually resident outside England and Wales. These events are, however, included in the totals for England and Wales combined, and for the United Kingdom.

From 1981 births to non-resident mothers in Northern Ireland are excluded from the figures for Northern Ireland, and for the United Kingdom.

Birth and death rates for 2004 have been calculated using the revised mid-2004 population estimates published on 20 December 2005.
Birth rates for 2005 are based on the 2004-based population projections for 2005.

Marriage and divorce rates in England and Wales for 1986 have been calculated using the interim revised marital status estimates (based on the original mid-2001 estimates) and are subject to further revision. Marriage and divorce rates for 2005 are based on 2004 marital status estimates.

Some stillbirths in 2004 are excluded from these and previously published figures, as the relevant registration details were not sent to ONS before the statistics were compiled. Revised figures for 2004 will be published as soon as possible to include the additional stillbirth registrations. See 'In brief' *Population Trends* 124.

See 'Notes to tables'.

Table 2.1 continued

Vital statistics summary

Constituent countries of the United Kingdom

Numbers (thousands) and rates

Year and quarter	All live births Number	Rate[1]	Live births outside marriage Number	Rate[2]	Marriages Number	Rate[3]	Divorces Number	Rate[4]	Deaths Number	Rate[1]	Infant mortality[5] Number	Rate[2]	Neonatal mortality[6] Number	Rate[2]	Perinatal mortality[7] Number	Rate[8]
Wales																
1976	33.4	11.9	2.9	86	19.5	..	..	..	36.3	13.0	0.46	13.7	0.32	9.6	0.64	19.0
1981	35.8	12.7	4.0	112	19.8	..	..	..	35.0	12.4	0.45	12.6	0.29	8.1	0.51	14.1
1986	37.0	13.1	7.8	211	19.5	..	7.9	..	34.7	12.3	0.35	9.5	0.21	5.6	0.38	10.3
1991	38.1	13.3	12.3	323	16.6	..	8.6	..	34.1	11.9	0.25	6.6	0.16	4.1	0.30	7.9
1996	34.9	12.1	14.4	412	14.8	..	8.4	..	34.6	12.0	0.20	5.6	0.13	3.6	0.26	7.5
1999	32.1	11.1	14.8	461	14.0	..	7.5	..	35.0	12.1	0.20	6.1	0.13	4.0	0.25	7.7
2000	31.3	10.8	14.8	472	14.1	..	7.2	..	33.3	11.5	0.17	5.3	0.11	3.5	0.23	7.2
2001	30.6	10.5	14.8	483	13.0	..	7.4	..	33.0	11.3	0.16	5.4	0.11	3.5	0.23	7.5
2002	30.2	10.3	15.0	497	13.5	..	7.6	..	33.2	11.3	0.14	4.5	0.10	3.2	0.24	7.7
2003	31.4	10.7	15.8	503	14.5	..	7.7	..	33.7	11.5	0.13	4.3	0.10	3.1	0.24	7.5
2004	32.3	10.9	16.6	513	14.8[P]	..	7.9[P]		32.1	10.9	0.16	4.9	0.10	3.1	0.25	7.8
2005	32.6	11.0[P]	17.1	524	..	..	7.1[P]	..	32.2[P]	10.9[P]	0.14[P]	4.3[P]	0.09[P]	2.9[P]	..	..
2003 March	7.5	10.3	3.8	505	1.7	..	2.0	..	8.9	12.3	0.04	4.7	0.03	3.8	0.06	7.7
June	7.8	10.7	3.9	494	4.0	..	2.0	..	8.3	11.4	0.03	4.0	0.02	2.7	0.06	7.3
Sept	8.3	11.2	4.2	503	6.2	..	2.0	..	7.6	10.2	0.04	4.6	0.03	3.5	0.07	8.2
Dec	7.8	10.5	4.0	511	2.6	..	1.8	..	8.8	11.9	0.03	3.8	0.02	2.3	0.05	6.9
2004 March	7.8	10.6	4.0	514	1.7[P]	..	2.0[P]	..	8.8	12.0	0.05	5.9	0.03	3.9	0.08	9.8
June	7.8	10.6	3.9	500	4.0[P]	..	2.0[P]	..	7.6	10.4	0.04	4.9	0.02	3.1	0.06	7.4
Sept	8.4	11.4	4.3	512	6.4[P]	..	2.1[P]	..	7.5	10.1	0.04	4.9	0.03	3.7	0.06	7.5
Dec	8.3	11.2	4.4	523	2.7[P]	..	1.8[P]	..	8.1	11.0	0.03	3.8	0.02	1.8	0.05	6.5
2005 March	7.8[P]	10.7[P]	4.1[P]	529[P]	..	..	1.8[P]	..	9.2[P]	12.6[P]	0.03[P]	3.8[P]	0.02[P]	2.9[P]	0.05[P]	6.9[P]
June	7.9[P]	10.7[P]	4.0[P]	510[P]	..	..	1.8[P]	..	8.0[P]	10.8[P]	0.04[P]	4.6[P]	0.03[P]	3.2[P]	0.06[P]	7.7[P]
Sept	8.7[P]	11.6[P]	4.6[P]	530[P]	..	..	1.8[P]	..	7.2[P]	9.6[P]	0.03[P]	3.8[P]	0.03[P]	2.9[P]	0.06[P]	6.7[P]
Dec	8.2	10.9[P]	4.3	527	..	..	1.7[P]	..	7.8[P]	10.5[P]	0.04[P]	4.9[P]	0.02[P]	2.6[P]	..	..
Scotland																
1976	64.9	12.5	6.0	93	37.5	53.8	8.1	6.5	65.3	12.5	0.96	14.8	0.67	10.3	1.20	18.3
1981	69.1	13.4	8.5	122	36.2	47.5	9.9	8.0	63.8	12.3	0.78	11.3	0.47	6.9	0.81	11.6
1986	65.8	12.9	13.6	206	35.8	42.9	12.8	10.7	63.5	12.4	0.58	8.8	0.34	5.2	0.67	10.2
1991	67.0	13.2	19.5	291	33.8	39.0	12.4	10.6	61.0	12.0	0.47	7.1	0.29	4.6	0.58	8.6
1996	59.3	11.6	21.4	360	30.2	33.2	12.3	10.9	60.7	11.9	0.37	6.2	0.23	3.9	0.55	9.2
1999	55.1	10.9	22.7	412	29.9	31.5	11.9	10.9	60.3	11.9	0.28	5.0	0.18	3.3	0.42	7.6
2000	53.1	10.5	22.6	426	30.4	31.6	11.1	10.3	57.8	11.4	0.31	5.7	0.21	4.0	0.45	8.4
2001	52.5	10.4	22.8	433	29.6	31.0	10.6	9.7	57.4	11.3	0.29	5.5	0.20	3.8	0.45	8.5
2002	51.3	10.1	22.5	440	29.8	30.8	10.8	10.0	58.1	11.5	0.27	5.3	0.16	3.2	0.39	7.6
2003	52.4	10.4	23.9	455	30.8	31.3	10.1	10.2	58.5	11.6	0.27	5.1	0.18	3.4	0.42	8.0
2004	54.0	10.6	25.2	467	32.2	32.2	11.2	10.5	56.2	11.1	0.27	4.9	0.17	3.1	0.44	8.1
2005	54.4[P]	10.7[P]	25.6[P]	471[P]	..	..	..	..	55.7[P]	10.9[P]	0.28[P]	5.2[P]	0.19[P]	3.5[P]	0.42[P]	7.7[P]
2003 March	12.8	10.3	5.9	462	3.7	15.2	2.5	9.4	15.7	12.6	0.07	5.5	0.05	3.8	0.09	6.9
June	12.9	10.3	5.8	447	8.4	34.2	3.0	11.1	14.1	11.2	0.06	4.3	0.03	2.5	0.11	8.2
Sept	13.8	10.8	6.2	448	12.3	49.7	2.6	9.7	13.3	10.4	0.07	4.9	0.05	3.4	0.11	8.1
Dec	13.0	10.2	6.0	464	6.4	25.5	2.7	10.1	15.4	12.1	0.07	5.6	0.05	3.8	0.12	8.9
2004 March	13.5	10.7	6.4	472	3.9	15.6	2.9	10.9	15.3	12.2	0.06	4.6	0.04	2.7	0.13	9.2
June	13.3	10.5	6.1	459	8.7	35.1	2.8	10.5	13.6	10.7	0.07	5.1	0.05	3.6	0.11	8.4
Sept	13.8	10.8	6.4	462	12.7	50.6	2.7	10.2	13.1	10.2	0.07	5.3	0.05	3.4	0.11	7.8
Dec	13.3	10.4	6.3	475	6.8	27.3	2.8	10.4	14.2	11.1	0.06	4.7	0.03	2.6	0.09	6.9
2005 March	13.4[P]	10.6[P]	6.2[P]	464[P]	3.8[P]	15.4[P]	2.6[P]	9.9[P]	15.6[P]	12.4[P]	0.07[P]	5.0[P]	0.04[P]	3.3[P]	0.09[P]	7.0[P]
June	13.6[P]	10.7[P]	6.4[P]	472[P]	8.6[P]	34.7[P]	2.8[P]	10.6[P]	13.7[P]	10.8[P]	0.07[P]	5.1[P]	0.05[P]	3.4[P]	0.13[P]	9.2[P]
Sept	14.2[P]	10.4[P]	6.7[P]	471[P]	12.3[P]	49.1[P]	2.7[P]	9.9[P]	12.8[P]	10.0[P]	0.08[P]	5.6[P]	0.06[P]	3.9[P]	0.11[P]	7.6[P]
Dec	13.2[P]	10.4[P]	6.3	477[P]	..	..	..	..	13.6[P]	10.6[P]	0.07[P]	5.2[P]	0.05[P]	3.4[P]	0.10[P]	7.1[P]
Northern Ireland																
1976	26.4	17.3	1.3	50	9.9	..	0.6	..	17.0	11.2	0.48	18.3	0.35	13.3	0.59	22.3
1981	27.2	17.6	1.9	70	9.6	45.4	1.4	4.2	16.3	10.6	0.36	13.2	0.23	8.3	0.42	15.3
1986	28.0	17.8	3.6	128	10.2	..	1.5	..	16.1	10.3	0.36	13.2	0.23	8.3	0.42	15.3
1991	26.0	16.2	5.3	203	9.2	..	2.3	..	15.1	9.4	0.19	7.4	0.12	4.6	0.22	8.4
1996	24.4	14.7	6.3	260	8.3	..	2.3	..	15.2	9.2	0.14	5.8	0.09	3.7	0.23	9.4
1999	23.0	13.7	7.0	303	7.6	..	2.3	..	15.7	9.3	0.15	6.4	0.11	4.8	0.23	10.0
2000	21.5	12.8	6.8	318	7.6	..	2.4	..	14.9	8.9	0.11	5.1	0.08	3.8	0.15	7.3
2001	22.0	13.0	7.1	325	7.3	..	2.4	..	14.5	8.6	0.13	6.1	0.10	4.5	0.19	8.5
2002	21.4	12.6	7.2	335	7.6	..	2.2	..	14.6	8.6	0.10	4.7	0.07	3.5	0.19	8.9
2003	21.6	12.7	7.4	344	7.8	..	2.3	..	14.5	8.5	0.11	5.3	0.09	4.0	0.18	8.1
2004	22.3	13.0	7.7	345	8.3	..	2.5	..	14.4	8.4	0.12	5.5	0.08	3.7	0.18	8.2
2005	22.3[P]	13.0[P]	8.1[P]	363[P]	..	..	..	..	14.2[P]	8.3[P]	0.14[P]	6.3[P]	0.11[P]	5.1[P]	0.18[P]	8.1[P]
2003 March	5.4	12.7	1.8	344	0.8	..	6.6	..	3.9	9.2	0.03	5.0	0.02	3.7	0.04	7.8
June	5.4	12.7	1.8	331	2.2	..	5.4	..	3.4	8.1	0.02	4.3	0.02	3.0	0.04	7.2
Sept	5.6	13.0	1.9	341	3.3	..	5.6	..	3.5	8.1	0.04	6.3	0.03	4.5	0.04	7.8
Dec	5.3	12.4	1.9	359	1.4	..	5.6	..	3.7	8.6	0.03	5.6	0.03	4.9	0.05	9.7
2004 March	5.7	13.3	2.0	352	0.9	..	7.7	..	3.9	9.1	0.03	5.5	0.02	3.5	0.05	7.9
June	5.4	12.7	1.8	337	2.4	..	6.5	..	3.6	8.4	0.03	5.9	0.02	4.4	0.05	9.5
Sept	5.8	13.5	2.0	339	3.5	..	5.5	..	3.4	8.0	0.04	6.0	0.02	4.1	0.05	8.3
Dec	5.4	12.7	1.9	353	1.6	..	5.5	..	3.5	8.1	0.02	4.4	0.02	2.8	0.04	7.0
2005 March	5.5[P]	13.0[P]	2.0[P]	363[P]	..	..	..	..	3.8[P]	8.9[P]	0.03[P]	5.2[P]	0.02[P]	4.3[P]	0.05[P]	8.8[P]
June	5.7[P]	13.3[P]	2.0[P]	359[P]	..	..	..	..	3.7[P]	8.5[P]	0.04[P]	7.2[P]	0.03[P]	5.6[P]	0.04[P]	8.4[P]
Sept	5.9[P]	13.7[P]	2.0[P]	358[P]	..	..	..	..	3.4[P]	7.7[P]	0.04[P]	6.6[P]	0.03[P]	5.6[P]	0.04[P]	7.2[P]
Dec	5.2[P]	11.9[P]	1.9[P]	373[P]	..	..	..	..	3.4[P]	7.8[P]	0.03[P]	6.0[P]	0.02[P]	4.6[P]	0.04[P]	7.9[P]

See notes opposite.

1 Per 1,000 population of all ages.
2 Per 1,000 live births.
3 Persons marrying per 1,000 unmarried population 16 and over.
4 Persons divorcing per 1,000 married population.
5 Deaths under 1 year.
6 Deaths under 4 weeks.
7 Stillbirths and deaths under 1 week. In October 1992 the legal definition of a stillbirth was changed, from baby born dead after 28 completed weeks of gestation or more, to one born dead after 24 completed weeks of gestation or more.
8 Per 1,000 live births and stillbirths.
p provisional.

Table 2.2	Key demographic and health indicators

Constituent countries of the United Kingdom

Numbers (thousands), rates, percentages, mean age

| | Population | Live births | Deaths | Dependency ratio | | Live births | | | | Expectation of life (in years) at birth | | | |
				Children[1]	Elderly[2]	TFR[3]	Standardised mean age of mother at birth (years)[4]	Unstandardised mean age of mother at birth (years)[5]	Outside marriage as percentage of total live births	Age-standardised mortality rate[6]	Males	Females	Infant mortality rate[7]
United Kingdom													
1976	56,216.1	675.5	680.8	42.1	29.5	1.74	..	26.4	9.0	10,486	..	..	14.5
1981	56,357.5	730.7	658.0	37.1	29.7	1.82	27.0	26.8	12.5	9,506	70.8	76.8	11.2
1986	56,683.8	754.8	660.7	33.5	29.7	1.78	27.4	27.0	21.4	8,914	71.9	77.7	9.5
1991	57,438.7	792.3	646.2	33.2	30.0	1.82	27.7	27.7	29.8	8,168	73.2	78.7	7.4
1996[8]	58,164.4	733.2	636.0	33.9	30.0	1.73	28.2	28.6	35.5	7,584	74.3	79.4	6.1
1999[8]	58,684.4	700.0	632.1	33.4	29.9	1.68	28.4	28.9	38.8	7,318	75.0	79.9	5.8
2000[8]	58,886.1	679.0	608.4	33.1	29.9	1.64	28.5	29.1	39.5	6,974	75.4	80.2	5.6
2001[8]	59,113.5	669.1	602.3	32.6	29.8	1.63	28.6	29.2	40.1	6,807	75.7	80.4	5.5
2002[8]	59,321.7	668.8	606.2	32.2	29.8	1.64	28.7	29.3	40.6	6,765	75.9	80.5	5.2
2003	59,553.8	695.6	612.0	31.8	29.9	1.71	28.8	29.4	41.5	6,757	76.3	80.7	5.3
2004[9]	59,834.3	716.0	583.1	31.4	30.0	1.77	28.9	29.4	42.3	6,390	..	..	5.0
2005[P]	..	722.6[P]	..	..	..	1.79[10][P]	29.1	29.5	..	6,278[10]	..	..	5.1
England													
1976	46,659.9	550.4	560.3	41.4	29.7	1.70	..	26.4	9.2	10,271	..	..	14.2
1981	46,820.8	598.2	541.0	36.4	29.9	1.79	..	26.8	12.9	9,298	71.1	77.0	10.9
1986	47,187.6	623.6	544.5	33.1	29.8	1.76	27.4	27.0	21.4	8,725	72.2	77.9	9.5
1991	47,875.0	660.8	534.0	32.9	30.0	1.81	27.7	27.7	30.1	8,017	73.4	78.9	7.3
1996[8]	48,519.1	614.2	524.0	33.7	30.0	1.73	28.2	28.7	35.5	7,414	74.5	79.6	6.1
1999[8]	49,032.9	589.5	519.6	33.3	29.9	1.69	28.4	29.0	38.5	7,138	75.3	80.1	5.7
2000[8]	49,233.3	572.8	501.0	33.0	29.8	1.65	28.5	29.2	39.1	6,821	75.7	80.4	5.6
2001[8]	49,449.7	563.7	496.1	32.5	29.7	1.63	28.6	29.3	39.6	6,650	76.0	80.6	5.4
2002[8]	49,646.9	565.7	499.1	32.1	29.7	1.65	28.7	29.4	40.1	6,603	76.2	80.7	5.2
2003	49,855.7	589.9	503.4	31.8	29.8	1.73	28.9	29.4	40.9	6,602	76.6	80.9	5.3
2004[9]	50,093.1	607.2	479.2	31.4	29.9	1.78	29.0	29.5	41.7	6,232	..	..	5.0
2005[P]	..	613.0	..	..	..	1.80[10][P]	29.1	29.5	..	6,125[10]	..	..	5.0
Wales													
1976	2,799.3	33.4	36.3	42.0	30.9	1.78	..	26.0	8.7	10,858	..	..	13.7
1981	2,813.5	35.8	35.0	37.6	31.6	1.87	..	26.6	11.2	9,846	70.4	76.4	12.6
1986	2,810.9	37.0	34.7	34.3	32.5	1.86	26.9	26.5	21.1	9,043	71.6	77.5	9.5
1991	2,873.0	38.1	34.1	34.4	33.5	1.88	27.1	27.0	32.3	8,149	73.1	78.8	6.6
1996[8]	2,891.3	34.9	34.6	34.9	33.7	1.81	27.5	27.8	41.2	7,758	73.9	79.1	5.6
1999[8]	2,900.6	32.1	35.0	34.4	33.6	1.72	27.6	28.1	46.1	7,637	74.7	79.6	6.1
2000[8]	2,906.9	31.3	33.3	34.1	33.5	1.68	27.7	28.2	47.2	7,180	74.9	79.8	5.3
2001[8]	2,910.2	30.6	33.0	33.7	33.6	1.66	27.8	28.3	48.3	7,017	75.4	80.1	5.4
2002[8]	2,923.4	30.2	33.2	33.2	33.6	1.63	28.0	28.4	49.7	6,951	75.7	80.2	4.5
2003	2,938.0	31.4	33.7	32.7	33.7	1.71	28.1	28.5	50.3	6,980	76.0	80.4	4.3
2004	2,952.5	32.1	32.1	32.2	33.9	1.77	28.2	28.5	51.3	6,582	..	..	4.9
2005[P]	..	32.6	..	..	..	1.79[10][P]	28.4	28.5	..	6,464[10]	..	..	4.3
Scotland													
1976	5,233.4	64.9	65.3	44.7	28.4	1.79	..	26.0	9.3	11,675	..	..	14.8
1981	5,180.2	69.1	63.8	38.2	28.4	1.84	..	26.3	12.2	10,849	69.1	75.3	11.3
1986	5,111.8	65.8	63.5	33.6	28.1	1.67	27.1	26.6	20.6	10,120	70.2	76.2	8.8
1991	5,083.3	67.0	61.0	32.4	28.9	1.69	27.5	27.4	29.1	9,216	71.4	77.1	7.1
1996	5,092.2	59.3	60.7	32.3	29.2	1.56	28.0	28.5	36.0	8,791	72.2	77.9	6.2
1999	5,072.0	55.1	60.3	31.7	29.7	1.51	28.3	28.9	41.2	8,493	72.8	78.4	5.0
2000	5,062.9	53.1	57.8	31.4	29.8	1.48	28.4	29.0	42.6	8,082	73.1	78.6	5.7
2001	5,064.2	52.5	57.4	30.8	30.0	1.49	28.5	29.2	43.3	7,930	73.3	78.8	5.5
2002	5,054.8	51.3	58.1	30.3	30.2	1.48	28.6	29.2	44.0	7,955	73.5	78.9	5.3
2003	5,057.4	52.4	58.5	29.9	30.3	1.54	28.7	29.3	45.5	7,922	73.8	79.1	5.1
2004	5,078.4	54.0	56.2	29.5	30.5	1.60	28.9	29.4	46.7	7,536	..	..	4.9
2005[P]	..	54.4[P]	..	..	..	1.62[10][P]	29.0	29.5	..	7,406[10]	..	..	5.2
Northern Ireland													
1976	1,523.5	26.4	17.0	56.1	25.3	2.68	..	27.4	5.0	11,746	..	..	18.3
1981	1,543.0	27.2	16.3	50.6	25.3	2.59	28.1	27.5	7.0	10,567	69.2	75.5	13.2
1986	1,573.5	28.0	16.1	46.1	25.5	2.45	28.1	27.5	12.8	10,071	70.9	77.1	13.2
1991	1,607.3	26.0	15.1	44.1	26.1	2.16	28.3	28.0	20.3	8,303	72.6	78.4	7.4
1996	1,661.8	24.4	15.2	41.8	25.5	1.95	28.7	28.8	26.0	7,742	73.8	79.2	5.8
1999	1,679.0	23.0	15.7	40.2	25.5	1.86	28.8	29.0	30.3	7,699	74.5	79.6	6.4
2000	1,682.9	21.5	14.9	39.5	25.4	1.75	29.0	29.2	31.8	7,279	74.8	79.8	5.1
2001	1,689.3	22.0	14.5	38.6	25.5	1.80	29.1	29.4	32.5	6,976	75.2	80.1	6.1
2002	1,696.6	21.4	14.6	37.9	25.7	1.77	29.2	29.5	33.5	6,930	75.6	80.4	4.7
2003	1,702.6	21.6	14.5	37.2	25.9	1.81	29.2	29.5	34.4	6,744	75.8	80.6	5.3
2004	1,710.3	22.3	14.4	36.4	26.2	1.87	29.4	29.7	34.5	6,609	..	..	5.5
2005[P]	..	22.3[P]	..	..	..	1.87[10][P]	29.5	29.7	..	6,411[10]	..	..	6.3

Notes: Some of these indicators are also in other tables. They are brought together to make comparison easier.

Figures for England and Wales represent the number of deaths registered in each year up to 1992, and the number of deaths occurring in each year from 1993 to 2004. Births and deaths figures for England and also for Wales exclude events for persons usually resident outside England and Wales. These events are, however, included in totals for England and Wales combined, and for the United Kingdom.

From 1981 births to non-resident mothers in Northern Ireland are excluded from the figures for Northern Ireland, and the United Kingdom.

1 Percentage of children under 16 to working population (males 16–64 and females 16–59).

2 Percentage of males 65 and over and females 60 and over to working population (males 16–64 and females 16–59).

3 TFR (total fertility rate) is the number of children that would be born to a woman if current patterns of fertility persisted throughout her childbearing life. It is sometimes called the TPFR (total period fertility rate).

4 Standardised to take account of the age structure of the population.

5 Unstandardised and therefore takes no account of the age structure of the population.

6 Per million population. The age-standardised mortality rate makes allowances for changes in the age structure of the population. See Notes to tables.

7 Deaths under one year per 1,000 live births.

8 These revised population estimates were published on 9 September 2004 (for mid-2001 and mid-2002) and 7 October (for mid-1992 to mid-2000), following the local authority population studies, and replace all earlier versions. All figures shown on this table are now therefore on a consistent basis.

9 Population estimates for mid-2004 for the United Kingdom, Great Britain, England and Wales and England were revised due to the Harrow correction that was published on 20 December 2005. Rates for 2004 have been calculated using these revised estimates.

10 Calculated using the 2004-based population projections for 2005.

p provisional

Table 3.1 — Live births: age of mother

England and Wales

Numbers (thousands), rates, mean age and TFRs

Year and quarter	Age of mother at birth							Mean[1] age (years)	Age of mother at birth							Mean[2] age (years)	TFR[3]
	All ages	Under 20	20–24	25–29	30–34	35–39	40 and over		All ages	Under 20	20–24	25–29	30–34	35–39	40 and over		
	Total live births (numbers)								Age-specific fertility rates[4]								
1961	811.3	59.8	249.8	248.5	152.3	77.5	23.3	27.6	89.2	37.3	172.6	176.9	103.1	48.1	15.0	27.4	2.77
1964(max)	876.0	76.7	276.1	270.7	153.5	75.4	23.6	27.2	92.9	42.5	181.6	187.3	107.7	49.8	13.7	27.3	2.93
1966	849.8	86.7	285.8	253.7	136.4	67.0	20.1	26.8	90.5	47.7	176.0	174.0	97.3	45.3	12.5	27.1	2.75
1971	783.2	82.6	285.7	247.2	109.6	45.2	12.7	26.2	83.5	50.6	152.9	153.2	77.1	32.8	8.7	26.6	2.37
1976	584.3	57.9	182.2	220.7	90.8	26.1	6.5	26.4	60.4	32.2	109.3	118.7	57.2	18.6	4.8	26.5	1.71
1977(min)	569.3	54.5	174.5	207.9	100.8	25.5	6.0	26.5	58.1	29.4	103.7	117.5	58.6	18.2	4.4	26.6	1.66
1981	634.5	56.6	194.5	215.8	126.6	34.2	6.9	26.8	61.3	28.1	105.3	129.1	68.6	21.7	4.9	27.0	1.80
1986	661.0	57.4	192.1	229.0	129.5	45.5	7.6	27.0	60.6	30.1	92.7	123.8	78.0	24.6	4.8	27.4	1.77
1991	699.2	52.4	173.4	248.7	161.3	53.6	9.8	27.7	63.6	33.0	89.3	119.4	86.7	32.1	5.3	27.7	1.82
1992	689.7	47.9	163.3	244.8	166.8	56.7	10.2	27.9	63.6	31.7	86.1	117.6	87.4	33.4	5.8	27.8	1.80
1993	673.5	45.1	152.0	236.0	171.1	58.8	10.5	28.1	62.7	30.9	82.5	114.4	87.4	34.1	6.2	27.9	1.76
1994	664.7	42.0	140.2	229.1	179.6	63.1	10.7	28.4	62.0	28.9	79.0	112.2	89.4	35.8	6.4	28.1	1.75
1995	648.1	41.9	130.7	217.4	181.2	65.5	11.3	28.5	60.5	28.5	76.4	108.4	88.3	36.3	6.8	28.2	1.72
1996	649.5	44.7	125.7	211.1	186.4	69.5	12.1	28.6	60.6	29.7	77.0	106.6	89.8	37.5	7.2	28.2	1.74
1997	643.1	46.4	118.6	202.8	187.5	74.9	12.9	28.8	60.0	30.2	76.0	104.3	89.8	39.4	7.6	28.3	1.73
1998	635.9	48.3	113.5	193.1	188.5	78.9	13.6	28.9	59.2	30.9	74.9	101.5	90.6	40.4	7.9	28.3	1.72
1999	621.9	48.4	110.7	181.9	185.3	81.3	14.3	29.0	57.8	30.9	73.0	98.3	89.6	40.6	8.1	28.4	1.70
2000	604.4	45.8	107.7	170.7	180.1	85.0	15.1	29.1	55.9	29.3	70.0	94.3	87.9	41.4	8.3	28.5	1.65
2001	594.6	44.2	108.8	159.9	178.9	86.5	16.3	29.2	54.7	28.0	69.0	91.7	88.0	41.5	8.8	28.6	1.63
2002	596.1	43.5	110.9	153.4	180.5	90.5	17.3	29.3	54.7	27.0	69.2	91.6	89.8	43.0	9.1	28.7	1.65
2003	621.5	44.2	116.6	156.9	187.2	97.4	19.1	29.4	56.8	26.8	71.2	96.4	94.8	46.4	9.8	28.8	1.73
2004[5]	639.7	45.1	121.1	160.0	190.6	102.2	20.8	29.4	58.2	26.9	72.7	98.4	99.4	48.9	10.4	28.9	1.78
2005[6]	645.8	45.0	122.1	164.3	188.2	104.1	22.2	29.5	58.5p	26.5p	71.9p	98.8p	100.9	50.3p	10.8p	29.0	1.80p
2000 March	148.7	11.4	26.4	42.5	44.1	20.6	3.6	29.1	55.3	29	69	95	87	40	8	28.5	1.64
June	150.7	11.1	26.0	42.8	45.7	21.4	3.7	29.2	56.0	29	68	95	90	42	8	28.6	1.66
Sept	155.0	11.8	27.8	43.6	46.2	21.7	3.9	29.1	57.0	30	72	96	90	42	9	28.5	1.69
Dec	150.1	11.5	27.5	41.8	44.1	21.4	3.9	29.1	55.2	29	71	92	86	41	9	28.5	1.64
2001 March	145.5	11.0	26.5	39.8	43.3	21.0	4.0	29.2	54.3	28	68	93	86	41	9	28.6	1.62
June	148.8	10.8	26.4	40.3	45.5	21.7	4.0	29.3	54.9	27	67	93	90	42	9	28.7	1.64
Sept	153.0	11.4	28.1	41.0	46.4	22.0	4.1	29.2	55.8	29	71	93	91	42	9	28.6	1.67
Dec	147.4	11.1	27.8	38.9	43.7	21.8	4.2	29.2	53.8	28	70	88	85	42	9	28.6	1.61
2002 March	143.3	10.5	26.5	37.4	43.2	21.6	4.1	29.3	53.3	26	67	91	87	42	9	28.7	1.61
June	147.2	10.4	26.7	37.9	45.5	22.4	4.3	29.4	54.1	26	67	91	91	43	9	28.8	1.63
Sept	155.0	11.4	28.9	39.9	46.9	23.4	4.5	29.3	56.4	28	72	95	93	44	9	28.7	1.70
Dec	150.6	11.2	28.8	38.2	45.0	23.0	4.5	29.3	54.8	28	71	91	89	44	9	28.7	1.65
2003 March	147.4	10.9	27.9	37.5	44.0	22.6	4.6	29.3	54.6	27	69	93	90	44	10	28.8	1.66
June	155.1	10.7	28.5	39.3	47.4	24.5	4.7	29.5	56.9	26	70	97	96	47	10	28.9	1.73
Sept	162.8	11.5	30.5	41.0	49.3	25.6	5.0	29.4	59.0	28	74	100	99	48	10	28.9	1.79
Dec	156.0	11.2	29.7	39.1	46.5	24.6	4.8	29.4	56.6	27	72	95	94	47	10	28.8	1.72
2004 March[5]	155.2	11.0	29.3	38.7	46.6	24.7	4.9	29.4	56.8	27	71	96	98	47	10	28.9	1.74
June[5]	157.4	10.7	29.3	39.4	47.7	25.2	5.0	29.5	57.6	26	71	97	100	49	10	29.0	1.77
Sept[5]	165.4	11.7	31.4	41.6	49.0	26.3	5.4	29.4	59.9	28	75	102	102	50	11	28.9	1.84
Dec[5]	161.7	11.6	31.1	40.3	47.2	26.0	5.5	29.4	58.6	28	74	99	98	49	11	28.9	1.80
2005 March[6]	154.3	10.9	29.3	38.9	44.9	24.8	5.4	29.5	56.6p	26p	70p	95p	98p	49p	11p	29.0	1.74p
June[6]	159.7	10.7	29.6	40.3	47.5	26.2	5.4	29.5	58.0p	25p	70p	97p	102p	51p	11p	29.1	1.78p
Sept[6]	170.2	11.8	32.4	43.5	49.3	26.9	5.7	29.4	60.9p	28p	76p	104p	105p	52p	11p	29.0	1.88p
Dec[6]	161.7	11.3	30.7	41.4	46.3	26.3	5.7	29.4	58.1p	27p	72p	99p	99p	50p	11p	29.0	1.79p

Notes: The rates for women of all ages, under 20, and 40 and over are based upon the populations of women aged 15–44, 15–19, and 40–44 respectively.

1 Unstandardised and therefore takes no account of the age structure of the population.
2 Standardised to take account of the age structure of the population. This measure is more appropriate for use when analysing trends or making comparisons between different geographies.
3 TFR (total fertility rate) is the number of children that would be born to a woman if current patterns of fertility persisted throughout her childbearing life. It is sometimes called the TPFR (total period fertility rate). During the post Second World War period the TFR reached a maximum in 1964 and a minimum in 1977.
4 Births per 1,000 women in the age-group; all quarterly age-specific fertility rates are adjusted for days in the quarter. They are not adjusted for seasonality.
5 Rates for 2004 have been calculated using the revised mid-2004 population estimates published on 20 December 2005.
6 Birth rates for 2005 are based on the 2004-based population projections for 2005.
p provisional

Table 3.2   **Live births outside marriage: age of mother and type of registration**

England and Wales   Numbers (thousands), mean age and percentages

Year and quarter	All ages	Under 20	20–24	25–29	30–34	35–39	40 and over	Mean[1] age (years)	All ages	Under 20	20–24	25–29	30–34	35–39	40 and over	Joint Same[3] address	Joint Different[3] addresses	Sole
	Live births outside marriage (numbers)								Percentage of total live births in age-group							As a percentage of all births outside marriage		
1971	65.7	21.6	22.0	11.5	6.2	3.2	1.1	23.7	8.4	26.1	7.7	4.7	5.7	7.0	9.0	45.5		54.5
1976	53.8	19.8	16.6	9.7	4.7	2.3	0.7	23.3	9.2	34.2	9.1	4.4	5.2	8.6	10.1	51.0		49.0
1981	81.0	26.4	28.8	14.3	7.9	1.3	0.9	23.4	12.8	46.7	14.8	6.6	6.2	3.9	12.5	58.2		41.8
1986	141.3	39.6	54.1	27.7	13.1	5.7	1.1	23.8	21.4	69.0	28.2	12.1	10.1	12.6	14.7	46.6	19.6	33.8
1991	211.3	43.4	77.8	52.4	25.7	9.8	2.1	24.8	30.2	82.9	44.9	21.1	16.0	18.3	21.3	54.6	19.8	25.6
1992	215.2	40.1	77.1	55.9	28.9	10.9	2.3	25.2	31.2	83.7	47.2	22.8	17.3	19.3	22.9	55.4	20.7	23.9
1993	216.5	38.2	75.0	57.5	31.4	11.9	2.5	25.5	32.2	84.8	49.4	24.4	18.4	20.2	23.5	54.8	22.0	23.2
1994	215.5	35.9	71.0	58.5	34.0	13.4	2.7	25.8	32.4	85.5	50.6	25.5	18.9	21.2	25.2	57.5	19.8	22.7
1995	219.9	36.3	69.7	59.6	37.0	14.4	3.0	26.0	33.9	86.6	53.3	27.4	20.4	22.0	26.2	58.1	20.1	21.8
1996	232.7	39.3	71.1	62.3	40.5	16.2	3.2	26.1	35.8	88.0	56.5	29.5	21.7	23.4	26.7	58.1	19.9	21.9
1997	238.2	41.1	69.5	63.4	42.2	18.2	3.7	26.2	37.0	88.7	58.6	31.3	22.5	24.3	28.6	59.5	19.3	21.2
1998	240.6	43.0	67.8	62.4	43.9	19.6	3.9	26.3	37.8	89.1	59.7	32.3	23.3	24.8	29.0	60.9	18.3	20.8
1999	241.9	43.0	67.5	61.2	45.0	20.8	4.3	26.4	38.9	89.0	61.0	33.6	24.3	25.6	30.2	61.8	18.2	19.9
2000	238.6	41.1	67.5	59.1	43.9	22.3	4.7	26.5	39.5	89.7	62.6	34.6	24.4	26.2	31.0	62.7	18.2	19.2
2001	238.1	39.5	68.1	56.8	45.2	23.3	5.1	26.7	40.0	89.5	62.6	35.5	25.3	26.9	31.6	63.2	18.4	18.4
2002	242.0	38.9	70.2	55.8	46.4	25.1	5.6	26.8	40.6	89.5	63.3	36.4	25.7	27.7	32.2	63.7	18.5	17.8
2003	257.2	39.9	75.7	58.2	49.2	27.8	6.4	26.9	41.4	90.2	64.9	37.1	26.3	28.5	33.3	63.5	19.0	17.4
2004	269.7	41.0	79.8	61.4	50.7	29.7	7.1	27.0	42.2	91.0	65.9	38.4	26.6	29.0	34.0	63.6	19.6	16.8
2005	276.5	41.2	82.1	64.4	50.8	30.3	7.7	27.0	42.8	91.8	67.2	39.2	27.0	29.1	34.8	63.5	20.2	16.2
1997 March	58.6	10.2	17.4	15.7	10.2	4.2	0.9	26.1	37.0	88.7	58.4	31.1	22.4	23.9	28.7	58.4	19.5	22.0
June	58.9	10.1	17.1	15.5	10.6	4.7	0.9	26.3	36.1	89.1	58.0	30.1	22.0	24.3	28.4	59.6	19.4	21.0
Sept	61.4	10.5	17.9	16.5	10.9	4.7	0.9	26.2	37.3	88.8	58.9	31.8	22.7	24.4	27.8	59.9	18.9	21.2
Dec	59.3	10.4	17.2	15.7	10.4	4.6	0.9	26.2	37.8	88.3	59.2	32.2	23.0	24.8	29.3	60.0	19.2	20.7
1998 March	58.5	10.4	16.5	15.3	10.7	4.6	1.0	26.3	37.5	89.0	59.5	31.9	23.1	24.4	29.6	60.5	18.4	21.1
June	58.4	10.3	16.2	15.4	10.8	4.7	0.9	26.3	36.8	89.6	59.1	31.8	22.5	24.0	28.3	61.0	18.2	20.8
Sept	63.2	11.3	17.9	16.3	11.5	5.2	1.0	26.3	38.1	89.2	60.0	32.3	23.6	25.2	28.5	60.9	18.4	20.7
Dec	60.5	11.0	17.2	15.4	10.9	5.0	1.0	26.3	38.9	88.5	60.4	33.3	24.0	25.7	29.7	61.2	18.4	20.4
1999 March	59.0	10.8	16.4	15.0	10.9	5.0	1.0	26.3	38.8	89.7	60.5	33.4	24.1	25.4	29.5	61.4	18.2	20.4
June	59.8	10.5	16.5	15.3	11.2	5.2	1.1	26.5	38.0	89.2	60.6	33.0	23.4	25.3	31.3	61.6	18.2	20.1
Sept	62.9	11.1	17.7	16.0	11.7	5.4	1.1	26.4	39.3	88.7	61.7	34.1	24.7	25.6	29.3	62.2	18.1	19.6
Dec	60.2	10.6	17.0	14.9	11.1	5.3	1.1	26.4	39.5	88.4	61.2	34.0	24.8	26.2	30.8	62.0	18.4	19.5
2000 March	59.0	10.2	16.5	14.8	10.9	5.4	1.2	26.5	39.7	89.7	62.6	34.8	24.7	26.1	31.7	62.5	18.1	19.5
June	57.9	10.0	16.1	14.4	10.9	5.5	1.1	26.6	38.5	89.7	61.9	33.5	23.8	25.7	30.6	62.9	17.8	19.2
Sept	61.7	10.6	17.6	15.3	11.3	5.7	1.2	26.5	39.8	89.7	63.3	35.0	24.5	26.5	30.4	62.7	18.1	19.2
Dec	60.1	10.3	17.3	14.7	10.9	5.7	1.2	26.5	40.0	89.5	62.8	35.2	24.7	26.6	31.4	62.6	18.6	18.8
2001 March	58.0	9.9	16.7	13.9	10.8	5.7	1.1	26.5	39.8	90.4	63.0	34.9	24.8	26.9	28.0	62.5	18.7	18.8
June	58.1	9.6	16.3	14.1	11.2	5.7	1.3	26.7	39.1	89.0	61.5	34.9	24.5	26.4	32.2	63.3	18.6	18.6
Sept	61.8	10.2	17.6	14.7	12.0	6.0	1.3	26.7	40.4	89.5	62.6	35.9	25.8	27.2	32.2	63.5	18.4	18.2
Dec	60.2	9.9	17.5	14.1	11.3	5.9	1.4	26.7	40.9	89.2	63.1	36.4	25.9	27.2	33.9	63.4	18.6	18.0
2002 March	58.0	9.4	16.7	13.6	10.9	6.0	1.3	26.8	40.5	89.4	63.0	36.4	25.4	27.7	31.5	63.2	18.5	18.3
June	58.3	9.3	16.6	13.5	11.4	6.1	1.4	26.8	39.6	89.4	62.2	35.6	25.0	27.2	31.7	64.2	18.2	17.7
Sept	63.4	10.2	18.4	14.6	12.3	6.5	1.5	26.8	40.9	89.3	63.8	36.6	26.1	27.9	32.7	63.9	18.5	17.5
Dec	62.3	10.0	18.4	14.1	11.9	6.5	1.5	26.8	41.4	89.7	64.1	36.9	26.4	28.0	32.8	63.3	18.9	17.8
2003 March	61.0	9.8	18.0	13.9	11.6	6.3	1.5	26.8	41.4	90.1	64.5	37.0	26.9	29.1	33.3	63.0	18.9	18.1
June	62.8	9.6	18.3	14.2	12.2	6.9	1.6	27.0	40.5	90.0	64.0	36.2	25.7	28.3	33.7	64.0	18.5	17.4
Sept	67.6	10.3	20.0	15.3	13.0	7.3	1.7	26.9	41.5	90.2	65.6	38.3	26.4	28.6	33.3	63.7	19.3	18.0
Dec	65.8	10.2	19.5	14.9	12.5	7.3	1.6	26.9	42.2	90.4	65.6	38.0	27.7	29.5	32.9	63.3	19.4	17.4
2004 March	65.2	10.1	19.3	14.8	12.5	7.0	1.7	26.9	42.0	91.2	65.8	38.2	26.8	28.2	34.3	63.1	19.4	17.4
June	65.2	9.8	19.1	14.9	12.5	7.3	1.7	27.0	41.4	91.0	65.1	37.7	26.2	28.8	34.5	63.9	19.5	16.6
Sept	70.2	10.7	20.7	16.1	13.0	7.9	1.8	27.0	42.4	91.2	66.1	38.6	26.5	30.0	33.5	63.7	19.7	16.6
Dec	69.1	10.6	20.7	15.7	12.7	7.5	1.9	26.9	42.7	90.6	66.6	39.0	27.0	29.0	33.9	63.6	19.8	16.6
2005 March	66.3	10.0	19.6	15.2	12.2	7.3	1.9	27.0	43.0	92.0	67.0	39.0	27.1	29.6	35.2	63.1	20.3	16.6
June	66.6	9.8	19.7	15.4	12.6	7.4	1.8	27.0	41.7	91.2	66.5	38.1	26.4	28.1	33.3	63.7	19.8	16.5
Sept	73.5	10.9	22.1	17.2	13.4	7.8	2.0	27.0	43.3	92.1	68.1	39.7	27.2	29.1	35.6	63.8	20.3	15.9
Dec	69.9	10.4	20.7	16.5	12.6	7.7	2.0	27.0	43.3	92.1	67.4	39.9	27.3	29.5	34.8	63.5	20.3	16.2

1   The mean ages in this table are unstandardised and therefore take no account of the structure of the population by age or marital status.
2   Births outside marriage can be registered by both the mother and father (joint) or by the mother alone (sole).
3   Usual address(es) of parents.
p   provisional

Table 3.3 Live births: within marriage, within marriage to remarried women, age of mother and birth order[1]

England and Wales

Numbers (thousands) and mean age

Live births within marriage / Live births within marriage to remarried women

Year and quarter	All ages	Under 20	20–24	25–29	30–34	35–39	40 and over	Mean[2] age (years)	All ages	Under 20	20–24	25–29	30–34	35–39	40 and over	Mean[2] age (years)
1971	717.5	61.1	263.7	235.7	103.4	42.1	11.6	26.4	19.4	0.1	2.1	6.6	6.1	3.4	1.1	33.1
1976	530.5	38.1	165.6	211.0	86.1	23.9	5.8	26.6	26.7	0.1	2.9	10.5	8.7	3.6	1.0	30.4
1981	553.5	30.1	165.7	201.5	118.7	31.5	6.0	27.2	38.8	0.1	3.6	13.4	14.1	6.2	1.4	30.9
1986	519.7	17.8	138.0	201.3	116.4	39.8	6.4	27.9	41.7	0.0	2.6	13.2	15.4	8.7	1.7	31.7
1991	487.9	8.9	95.6	196.3	135.5	43.8	7.7	28.9	39.4	0.0	1.6	10.8	15.8	9.1	2.1	32.4
1995	428.2	5.6	61.0	157.9	144.2	51.1	8.4	29.8	33.3	0.0	0.8	7.2	14.0	9.1	2.1	33.2
1996	416.8	5.4	54.7	148.8	145.9	53.3	8.9	30.0	32.6	0.0	0.7	6.4	13.9	9.3	2.2	33.4
1997	404.9	5.2	49.1	139.4	145.3	56.7	9.2	30.3	31.4	0.0	0.6	5.8	13.1	9.5	2.4	33.6
1998	395.3	5.3	45.7	130.7	144.6	59.3	9.6	30.5	30.2	0.0	0.6	5.1	12.4	9.7	2.4	33.9
1999	380.0	5.3	43.2	120.7	140.3	60.5	9.9	30.6	27.5	0.0	0.4	4.3	11.3	9.1	2.4	34.1
2000	365.8	4.7	40.3	111.6	136.2	62.7	10.4	30.8	25.8	0.0	0.4	3.7	10.4	8.9	2.4	34.3
2001	356.5	4.6	40.7	103.1	133.7	63.2	11.1	30.9	23.9	0.0	0.4	3.1	9.5	8.6	2.4	34.5
2002	354.1	4.6	40.7	97.6	134.1	65.4	11.8	31.0	22.8	0.0	0.3	2.7	8.9	8.5	2.5	34.7
2003	364.2	4.3	40.9	98.7	138.0	69.6	12.7	31.2	22.6	0.0	0.3	2.4	8.4	8.8	2.6	35.0
2004	370.0	4.1	41.3	98.5	139.8	72.6	13.7	31.2	21.5	0.0	0.3	2.2	7.7	8.6	2.7	35.1
2005	369.3	3.7	40.0	100.0	137.4	73.8	14.5	31.3	20.0	0.0	0.3	2.1	6.8	8.1	2.7	35.3
2004 March	89.9	1.0	10.0	23.9	34.1	17.7	3.2	31.2	5.3	0.0	0.1	0.6	1.9	2.1	0.6	35.1
June	92.2	1.0	10.3	24.5	35.2	18.0	3.3	31.2	5.3	0.0	0.1	0.5	2.0	2.1	0.6	35.0
Sept	95.3	1.0	10.6	25.5	36.0	18.4	3.6	31.2	5.6	0.0	0.1	0.6	2.0	2.2	0.7	35.1
Dec	92.6	1.1	10.4	24.6	34.5	18.4	3.6	31.3	5.3	0.0	0.1	0.5	1.9	2.1	0.7	35.3
2005 March	88.0	0.9	9.7	23.7	32.8	17.4	3.5	31.3	4.9	0.0	0.1	0.5	1.7	1.9	0.6	35.3
June	93.2	0.9	9.9	24.9	35.0	18.8	3.6	31.3	5.0	0.0	0.1	0.5	1.7	2.0	0.7	35.2
Sept	96.2	0.9	10.4	26.2	35.9	19.1	3.7	31.3	5.3	0.0	0.1	0.6	1.8	2.1	0.7	35.3
Dec	91.7	0.9	10.0	24.9	33.7	18.5	3.7	31.3	5.0	0.0	0.1	0.5	1.7	2.0	0.7	35.3

First live births / Second live births

Year and quarter	All ages	Under 20	20–24	25–29	30–34	35–39	40 and over	Mean[2] age (years)	All ages	Under 20	20–24	25–29	30–34	35–39	40 and over	Mean[2] age (years)
1971	283.6	49.5	135.8	74.8	17.2	5.1	1.2	23.9	240.8	10.7	93.6	94.1	31.8	8.9	1.7	26.2
1976	217.2	30.2	85.4	77.2	19.7	3.9	0.7	24.8	203.6	7.4	62.5	91.8	34.7	6.2	1.0	26.8
1981	224.3	23.6	89.5	77.2	27.8	5.4	0.7	25.3	205.7	6.1	59.0	82.7	47.7	9.1	1.1	27.4
1986	206.9	13.8	74.7	79.3	30.8	7.5	0.9	26.2	189.2	3.6	47.5	78.9	45.5	12.3	1.3	28.0
1991	193.7	6.7	51.2	84.5	40.2	9.7	1.3	27.5	178.3	2.0	32.8	73.9	53.0	14.7	1.9	28.9
1995	168.1	4.3	32.3	71.0	46.6	12.1	1.8	28.5	158.1	1.2	20.6	57.3	58.5	18.1	2.4	30.0
1996	163.0	4.2	28.9	67.2	47.7	13.1	1.9	28.8	153.8	1.0	18.5	53.4	59.1	19.2	2.6	30.3
1997	157.0	4.1	25.9	63.1	48.1	13.8	2.0	29.0	150.4	1.0	16.6	50.0	59.4	20.7	2.7	30.5
1998	155.7	4.2	24.3	60.6	49.5	15.0	2.1	29.2	146.9	1.0	15.5	46.4	58.9	22.2	2.8	30.7
1999	153.4	4.3	23.5	57.4	50.0	16.1	2.2	29.3	139.5	0.9	14.4	41.8	56.6	22.6	3.1	30.9
2000	146.5	3.8	21.6	52.7	49.4	16.6	2.4	29.6	134.7	0.8	13.7	38.4	54.8	23.8	3.2	31.1
2001	143.9	3.8	22.2	48.8	49.7	16.8	2.6	29.6	132.2	0.8	13.7	35.7	53.8	24.8	3.5	31.2
2002	145.2	3.8	22.4	47.1	51.0	18.1	2.8	29.8	130.3	0.7	13.5	33.0	53.7	25.6	3.8	31.4
2003	151.0	3.5	22.2	48.4	54.2	19.6	3.1	29.9	132.9	0.8	13.9	32.5	54.3	27.1	4.2	31.5
2004	154.5	3.3	22.6	48.9	55.5	20.7	3.5	30.0	133.7	0.7	13.8	31.9	54.5	28.3	4.5	31.6
2005	156.0	3.0	22.1	50.0	55.7	21.4	3.8	30.1	132.0	0.6	13.2	32.1	52.8	28.6	4.8	31.7
2004 March	36.9	0.8	5.3	11.7	13.4	5.0	0.8	30.0	33.0	0.2	3.4	7.9	13.4	7.0	1.1	31.6
June	37.6	0.8	5.5	12.0	13.5	5.0	0.8	30.0	34.4	0.2	3.5	8.2	14.3	7.1	1.1	31.6
Sept	40.3	0.8	5.9	12.8	14.6	5.3	0.9	30.0	34.1	0.2	3.5	8.2	13.8	7.2	1.1	31.6
Dec	39.8	0.9	5.8	12.4	14.1	5.5	1.0	30.1	32.2	0.2	3.3	7.6	13.0	6.9	1.2	31.7
2005 March	36.4	0.7	5.2	11.6	13.0	5.0	0.9	30.1	31.9	0.1	3.3	7.8	12.7	6.8	1.1	31.6
June	38.6	0.8	5.5	12.4	13.7	5.3	1.0	30.1	34.1	0.2	3.3	8.2	14.0	7.4	1.2	31.7
Sept	40.8	0.8	5.7	13.2	14.6	5.6	0.9	30.1	34.2	0.2	3.4	8.3	13.7	7.4	1.2	31.7
Dec	39.9	0.7	5.6	12.7	14.2	5.6	1.0	30.2	31.8	0.2	3.2	7.7	12.5	7.0	1.2	31.7

Third live births / Fourth and higher order live births[3]

Year and quarter	All ages	Under 20	20–24	25–29	30–34	35–39	40 and over	Mean[2] age (years)	All ages	Under 20	20–24	25–29	30–34	35–39	40 and over	Mean[2] age (years)
1971	111.7	0.9	26.6	43.6	27.9	10.4	2.2	28.7	81.4	0.1	7.6	23.2	26.5	17.6	6.5	30.7
1976	71.0	0.5	14.4	29.8	19.5	5.8	1.1	28.8	38.8	0.0	3.3	12.2	12.1	8.0	3.1	30.7
1981	82.4	0.4	14.1	29.5	28.7	8.7	1.0	29.5	41.1	0.0	3.1	12.0	14.5	8.3	3.2	31.1
1986	80.8	0.3	12.7	30.2	25.6	10.5	1.5	29.9	42.7	0.0	3.1	13.0	14.5	9.4	2.8	31.2
1991	76.1	0.2	9.4	26.8	27.5	10.5	1.8	30.4	39.8	0.0	2.3	11.1	14.8	8.9	2.7	31.6
1995	66.7	0.1	6.5	20.5	26.1	11.7	1.8	31.1	35.3	0.0	1.6	9.0	13.1	9.2	2.4	32.0
1996	65.3	0.1	5.8	19.6	26.0	12.0	1.8	31.3	34.7	0.0	1.5	8.6	13.1	9.0	2.6	32.2
1997	63.2	0.1	5.3	18.1	25.1	12.7	2.0	31.5	34.2	0.0	1.4	8.1	12.7	9.4	2.6	32.4
1998	60.4	0.1	4.7	16.4	24.0	13.1	2.1	31.8	32.3	0.0	1.2	7.4	12.1	9.0	2.6	32.6
1999	56.4	0.1	4.2	14.7	22.3	13.0	2.1	32.0	30.7	0.0	1.1	6.8	11.4	8.8	2.6	32.7
2000	54.9	0.1	4.0	14.1	21.1	13.5	2.2	32.1	29.7	0.0	1.0	6.4	10.9	8.7	2.7	32.8
2001	52.1	0.1	3.9	12.8	19.8	13.2	2.3	32.2	28.3	0.0	0.9	5.9	10.4	8.4	2.7	33.0
2002	50.3	0.1	3.9	11.8	19.0	13.1	2.4	32.3	28.2	0.0	0.9	5.6	10.3	8.5	2.8	33.1
2003	52.0	0.1	3.8	12.1	19.2	14.1	2.6	32.5	28.4	0.0	1.0	5.7	10.2	8.8	2.8	33.1
2004	52.5	0.1	4.0	12.1	19.3	14.3	2.7	32.5	29.3	0.0	0.9	5.7	10.5	9.2	2.9	33.2
2005	52.2	0.1	3.8	12.3	18.7	14.5	2.9	32.5	29.2	0.0	0.9	5.6	10.2	9.4	3.0	33.3
2004 March	12.9	0.0	1.0	3.0	4.8	3.5	0.6	32.4	7.1	0.0	0.2	1.4	2.6	2.2	0.7	33.2
June	13.1	0.0	1.0	2.9	4.9	3.6	0.7	32.5	7.2	0.0	0.2	1.4	2.6	2.3	0.7	33.1
Sept	13.4	0.0	1.0	3.1	4.9	3.6	0.7	32.5	7.5	0.0	0.3	1.4	2.7	2.4	0.8	33.3
Dec	13.2	0.0	1.0	3.1	4.8	3.7	0.7	32.5	7.5	0.0	0.2	1.5	2.7	2.4	0.7	33.2
2005 March	12.6	0.0	0.9	3.0	4.5	3.4	0.7	32.5	7.1	0.0	0.2	1.3	2.5	2.3	0.8	33.4
June	13.1	0.0	0.9	3.0	4.7	3.6	0.7	32.6	7.3	0.0	0.2	1.3	2.6	2.4	0.7	33.4
Sept	13.7	0.0	1.0	3.3	4.9	3.8	0.7	32.5	7.5	0.0	0.2	1.5	2.7	2.3	0.8	33.2
Dec	12.8	0.0	0.9	3.0	4.5	3.6	0.7	32.6	7.2	0.0	0.2	1.4	2.4	2.4	0.8	33.3

1 Birth order is based on all live births within marriage to the mother by her present or any former husband.
2 The mean ages shown in this table are unstandardised and therefore take no account of changes in the structure of the population by age, marital status or parity.
3 Mean age at birth refers to fourth births only.
p provisional.

Table 4.1 — Conceptions: age of women at conception

England and Wales (residents)
Numbers (thousands) and rates; and percentage terminated by abortion

Year and quarter	All ages	Under 16	Under 18	Under 20	20–24	25–29	30–34	35–39	40 and over
(a) numbers (thousands)									
1991	853.7	7.5	40.1	101.6	233.3	281.5	167.5	57.6	12.1
1996	816.9	8.9	43.5	94.9	179.8	252.6	200.0	75.5	14.1
1999	774.0	7.9	42.0	98.8	157.6	218.5	197.1	86.0	16.0
2000	767.0	8.1	41.3	97.7	159.0	209.3	195.3	88.7	17.0
2001	763.7	7.9	41.0	96.0	161.6	199.3	196.7	92.2	17.8
2002	787.0	7.9	42.0	97.1	167.8	199.4	204.3	98.9	19.6
2003	806.8	8.0	42.2	98.6	175.3	199.8	209.0	103.1	20.9
2004[p]	825.8	7.6	42.2	101.2	181.1	204.9	209.9	106.6	22.7
2001 March	189.2	1.9	10.2	24.3	40.4	50.0	47.8	22.3	4.4
June	187.4	2.1	10.2	24.0	39.8	48.8	47.7	22.8	4.4
Sept	189.3	1.9	10.0	23.1	39.2	49.5	49.9	23.2	4.4
Dec	197.9	2.0	10.6	24.6	42.3	51.1	51.3	23.9	4.7
2002 March	191.6	1.9	10.3	24.1	41.3	48.8	49.0	23.7	4.6
June	190.4	2.0	10.5	24.2	40.7	48.2	48.8	23.8	4.8
Sept	197.4	2.0	10.2	23.4	41.4	50.2	52.4	25.2	4.9
Dec	207.6	2.0	11.0	25.4	44.4	52.3	54.2	26.2	5.2
2003 March	198.2	1.9	10.5	24.5	42.9	49.4	51.2	25.2	4.9
June	198.5	2.1	10.8	24.7	43.2	49.1	51.1	25.2	5.2
Sept	200.1	2.0	10.2	23.7	43.1	49.3	52.8	26.1	5.2
Dec	210.0	2.0	10.7	25.7	46.1	52.0	54.0	26.7	5.6
2004 March[p]	207.9	2.0	10.9	26.2	45.9	51.1	52.6	26.6	5.6
June[p]	200.0	1.9	10.6	25.0	43.7	49.3	50.4	25.9	5.7
Sept[p]	203.6	1.8	10.0	24.0	44.1	50.7	52.7	26.6	5.6
Dec[p]	214.3	1.9	10.8	26.0	47.5	53.8	53.7	27.5	5.8
2005 March[p]	204.2	1.9	10.4	25.0	45.3	50.7	51.0	26.5	5.7
(b) rates (conceptions per thousand women in age group)[1]									
1991	77.7	8.9	44.6	64.1	120.2	135.1	90.1	34.4	6.6
1996	76.2	9.5	46.3	63.2	110.1	127.6	96.3	40.7	8.4
1999	71.9	8.3	45.1	63.1	103.9	118.0	95.3	42.9	9.1
2000	70.9	8.3	43.9	62.5	103.2	115.7	95.3	43.2	9.4
2001	70.3	8.0	42.7	60.8	102.5	114.2	96.7	44.3	9.6
2002	72.2	7.9	42.8	60.3	104.6	119.1	101.6	47.0	10.3
2003	73.7	8.0	42.3	59.8	107.1	122.8	105.9	49.1	10.7
2004[p]	75.2	7.5	41.7	60.3	108.7	126.0	109.2	50.9	11.3
2001 March	70.7	7.8	43.3	62.7	104.8	114.5	95.0	43.7	9.7
June	69.2	8.4	42.8	61.0	101.4	111.6	94.0	44.0	9.5
Sept	69.1	7.7	41.1	57.8	98.4	113.1	97.6	44.2	9.3
Dec	72.1	8.1	43.5	61.4	105.6	118.0	100.5	45.4	10.0
2002 March	71.3	7.7	42.9	61.3	105.1	116.4	98.4	45.8	9.9
June	70.1	8.1	42.9	60.4	101.9	114.8	97.1	45.5	10.2
Sept	71.8	7.7	41.2	57.5	102.1	119.4	103.5	47.6	10.2
Dec	75.4	8.0	44.1	62.1	108.9	125.1	107.6	49.4	10.7
2003 March	73.5	7.8	42.8	60.8	107.2	121.8	104.5	48.6	10.3
June	72.8	8.3	43.3	60.3	106.1	120.6	103.5	48.0	10.8
Sept	72.5	7.9	40.5	56.8	104.2	120.2	106.4	49.3	10.5
Dec	76.0	7.8	42.5	61.4	110.9	126.8	109.7	50.5	11.2
2004 March[p]	76.2	7.8	43.4	63.1	111.5	126.3	109.1	51.1	11.4
June[p]	73.3	7.7	42.1	60.1	105.8	122.1	105.3	49.7	11.5
Sept[p]	73.7	7.1	39.2	56.8	105.0	123.6	109.9	50.6	11.1
Dec[p]	77.5	7.4	42.3	61.4	112.6	130.5	113.1	52.5	11.3
2005 March[p]	75.2	7.6	41.4	60.1	108.9	124.7	110.5	51.7	11.4
(c) percentage terminated by abortion									
1991	19.4	51.1	39.9	34.5	22.2	13.4	13.7	22.0	41.6
1996	20.8	49.2	40.0	36.2	25.7	15.6	14.1	21.2	37.6
1999	22.6	52.6	43.0	38.6	28.5	17.5	14.7	21.2	37.0
2000	22.7	54.0	44.2	39.3	29.2	17.7	14.5	20.5	35.4
2001	23.2	55.8	45.7	40.4	29.7	18.4	14.6	20.4	34.6
2002	22.5	55.6	45.3	39.9	28.8	17.9	13.9	19.5	34.6
2003	22.5	57.4	45.7	40.2	29.0	17.9	13.6	18.9	34.7
2004[p]	22.4	57.1	45.6	40.1	28.9	18.2	13.2	18.4	33.0
2001 March	23.4	54.4	44.9	40.2	29.8	18.6	14.8	20.7	34.9
June	23.8	58.8	47.0	41.1	30.3	18.6	15.3	21.0	36.0
Sept	22.5	55.0	45.7	40.1	29.2	18.1	13.8	19.9	33.5
Dec	22.9	54.9	45.2	40.0	29.5	18.1	14.4	20.2	34.1
2002 March	22.9	54.3	44.9	40.2	29.4	18.1	14.1	19.8	35.1
June	22.9	55.5	45.0	39.4	28.9	18.4	14.5	20.1	34.8
Sept	21.6	56.1	45.0	39.4	27.8	17.3	13.2	18.7	34.2
Dec	22.6	56.4	46.3	40.7	29.0	17.8	13.9	19.4	34.5
2003 March	22.8	58.9	46.1	40.2	29.5	17.9	13.8	19.7	34.5
June	23.1	58.3	46.2	40.9	29.3	18.4	14.2	19.2	36.1
Sept	21.6	56.9	45.3	39.5	28.0	17.1	13.0	18.0	33.8
Dec	22.5	55.7	45.0	40.3	29.0	18.1	13.5	18.5	34.5
2004 March[p]	22.7	58.2	45.7	40.2	29.4	18.5	13.4	18.2	32.9
June[p]	23.0	57.2	46.4	40.8	29.2	18.6	13.7	19.2	33.5
Sept[p]	21.9	56.8	45.8	40.1	28.4	17.9	12.8	17.8	33.0
Dec[p]	22.1	56.1	44.5	39.3	28.7	17.8	13.1	18.3	32.6
2005 March[p]	22.4	57.3	47.2	41.0	29.1	18.0	13.1	17.9	32.5

Notes: Conceptions are estimates derived from birth registrations and abortion notifications.

Rates for women of all ages, under 16, under 18, under 20 and 40 and over are based on the population of women aged 15–44, 13–15, 15–17, 15–19 and 40–44 respectively.

For a quarterly analysis of conceptions under 18 for local authority areas see the National Statisitcs website, www.statistics.gov.uk

1 Rates for provisional annual 2004 conceptions are based on mid-year population estimates published on 20 December 2005.

Rates for provisional 2005 conceptions are based on the 2004-based population projections for 2005.

p provisional

Table 5.1 — Expectation of life at birth and selected age

Constituent countries of the United Kingdom   Years

Year	Males At birth	5	20	30	50	60	70	80	Year	Females At birth	5	20	30	50	60	70	80
United Kingdom																	
1981	70.8	66.9	52.3	42.7	24.1	16.3	10.1	5.8	1981	76.8	72.7	57.9	48.2	29.2	20.8	13.3	7.5
1986	71.9	67.8	53.2	43.6	24.9	16.8	10.5	6.0	1986	77.7	73.4	58.6	48.8	29.8	21.2	13.8	7.8
1991	73.2	68.9	54.2	44.7	26.0	17.7	11.1	6.4	1991	78.7	74.3	59.5	49.7	30.6	21.9	14.3	8.2
1996	74.3	69.8	55.1	45.6	26.9	18.5	11.6	6.6	1996	79.4	74.9	60.1	50.3	31.2	22.3	14.5	8.3
1997	74.5	70.1	55.4	45.9	27.2	18.8	11.7	6.7	1997	79.6	75.1	60.2	50.4	31.3	22.5	14.6	8.4
1998	74.8	70.3	55.6	46.1	27.4	18.9	11.9	6.7	1998	79.7	75.2	60.4	50.5	31.4	22.6	14.7	8.4
1999	75.0	70.6	55.9	46.3	27.6	19.2	12.0	6.8	1999	79.9	75.4	60.5	50.7	31.6	22.8	14.8	8.5
2000	75.4	70.9	56.2	46.6	28.0	19.5	12.3	7.0	2000	80.2	75.6	60.8	51.0	31.9	23.0	15.0	8.6
2001	75.7	71.2	56.5	46.9	28.3	19.8	12.5	7.1	2001	80.4	75.9	61.0	51.2	32.1	23.2	15.2	8.7
2002	75.9	71.5	56.7	47.2	28.5	20.0	12.6	7.2	2002	80.5	76.0	61.1	51.3	32.2	23.3	15.2	8.7
2003	76.3	71.8	57.0	47.4	28.8	20.2	12.9	7.3	2003	80.7	76.2	61.3	51.5	32.4	23.4	15.3	8.7
England and Wales																	
1981	71.0	67.1	52.5	42.9	24.3	16.4	10.1	5.8	1981	77.0	72.9	58.1	48.3	29.4	20.9	13.4	7.5
1986	72.1	68.0	53.4	43.8	25.0	16.9	10.5	6.1	1986	77.9	73.6	58.8	49.0	30.0	21.4	13.9	7.9
1991	73.4	69.1	54.4	44.8	26.1	17.8	11.2	6.4	1991	78.9	74.5	59.7	49.9	30.8	22.0	14.4	8.3
1996	74.5	70.1	55.4	45.8	27.1	18.7	11.6	6.6	1996	79.6	75.1	60.2	50.4	31.3	22.5	14.6	8.4
1997	74.8	70.3	55.6	46.1	27.4	18.9	11.8	6.7	1997	79.7	75.2	60.4	50.6	31.5	22.6	14.7	8.4
1998	75.0	70.6	55.8	46.3	27.6	19.1	11.9	6.8	1998	79.9	75.4	60.5	50.7	31.6	22.7	14.8	8.4
1999	75.3	70.8	56.1	46.5	27.8	19.3	12.1	6.9	1999	80.1	75.6	60.7	50.9	31.8	22.9	14.9	8.5
2000	75.6	71.2	56.4	46.9	28.1	19.6	12.3	7.0	2000	80.3	75.8	61.0	51.1	32.0	23.1	15.1	8.6
2001	76.0	71.5	56.7	47.2	28.5	19.9	12.6	7.1	2001	80.6	76.0	61.2	51.4	32.2	23.3	15.2	8.7
2002	76.2	71.7	57.0	47.4	28.7	20.1	12.7	7.2	2002	80.7	76.1	61.3	51.5	32.3	23.4	15.3	8.7
2003	76.5	72.0	57.3	47.7	28.9	20.4	13.0	7.3	2003	80.9	76.4	61.5	51.7	32.5	23.6	15.4	8.8
England																	
1981	71.1	67.1	52.5	42.9	24.3	16.4	10.1	5.8	1981	77.0	72.9	58.2	48.4	29.4	20.9	13.4	7.5
1986	72.2	68.1	53.4	43.8	25.1	17.0	10.6	6.1	1986	77.9	73.6	58.8	49.0	30.0	21.4	13.9	7.9
1991	73.4	69.1	54.4	44.9	26.2	17.8	11.2	6.4	1991	78.9	74.5	59.7	49.9	30.8	22.0	14.4	8.3
1996	74.5	70.1	55.4	45.9	27.1	18.7	11.7	6.6	1996	79.6	75.1	60.3	50.5	31.3	22.5	14.6	8.4
1997	74.8	70.4	55.6	46.1	27.4	18.9	11.8	6.7	1997	79.8	75.3	60.4	50.6	31.5	22.6	14.7	8.4
1998	75.0	70.6	55.9	46.3	27.6	19.1	12.0	6.8	1998	79.9	75.4	60.6	50.7	31.6	22.7	14.8	8.5
1999	75.3	70.9	56.1	46.6	27.9	19.4	12.1	6.9	1999	80.1	75.6	60.8	50.9	31.8	22.9	14.9	8.5
2000	75.7	71.2	56.5	46.9	28.2	19.6	12.4	7.0	2000	80.4	75.8	61.0	51.2	32.0	23.1	15.1	8.6
2001	76.0	71.5	56.8	47.2	28.5	19.9	12.6	7.1	2001	80.6	76.1	61.2	51.4	32.3	23.4	15.3	8.7
2002	76.2	71.8	57.0	47.4	28.7	20.1	12.8	7.2	2002	80.7	76.2	61.3	51.5	32.4	23.4	15.3	8.7
2003	76.6	72.1	57.3	47.7	29.0	20.4	13.0	7.3	2003	80.9	76.4	61.5	51.7	32.6	23.6	15.5	8.8
Wales																	
1981	70.4	66.5	51.9	42.2	23.6	15.8	9.7	5.6	1981	76.4	72.3	57.5	47.7	28.9	20.5	13.1	7.4
1986	71.6	67.5	52.8	43.2	24.6	16.6	10.3	6.0	1986	77.5	73.3	58.5	48.7	29.7	21.1	13.7	7.8
1991	73.1	68.8	54.1	44.6	25.8	17.6	11.0	6.4	1991	78.8	74.3	59.5	49.7	30.6	21.8	14.3	8.3
1996	73.9	69.4	54.7	45.3	26.6	18.2	11.3	6.4	1996	79.1	74.6	59.7	49.9	30.9	22.1	14.4	8.3
1997	74.3	69.8	55.1	45.6	26.9	18.5	11.6	6.6	1997	79.3	74.8	60.0	50.2	31.1	22.3	14.5	8.4
1998	74.4	70.0	55.2	45.8	27.1	18.6	11.6	6.6	1998	79.4	74.9	60.0	50.2	31.1	22.3	14.5	8.3
1999	74.7	70.2	55.5	46.1	27.4	18.9	11.9	6.8	1999	79.6	75.1	60.2	50.4	31.3	22.5	14.6	8.4
2000	74.9	70.5	55.8	46.3	27.6	19.1	12.0	6.8	2000	79.8	75.3	60.4	50.6	31.5	22.6	14.7	8.4
2001	75.4	70.9	56.2	46.7	28.0	19.5	12.3	7.1	2001	80.1	75.5	60.6	50.8	31.8	22.9	14.9	8.5
2002	75.7	71.1	56.3	46.9	28.2	19.7	12.4	7.1	2002	80.2	75.6	60.7	50.9	31.8	22.9	15.0	8.6
2003	76.0	71.4	56.7	47.1	28.5	20.0	12.6	7.2	2003	80.4	75.8	60.9	51.1	32.0	23.1	15.1	8.6
Scotland																	
1981	69.1	65.2	50.6	41.1	22.9	15.4	9.6	5.5	1981	75.3	71.2	56.4	46.7	27.9	19.7	12.7	7.2
1986	70.2	66.0	51.4	41.9	23.5	15.8	9.9	5.7	1986	76.2	71.9	57.1	47.3	28.4	20.1	13.0	7.5
1991	71.4	67.1	52.5	43.0	24.6	16.6	10.4	6.1	1991	77.1	72.7	57.9	48.1	29.2	20.7	13.5	7.9
1996	72.2	67.8	53.1	43.7	25.3	17.3	10.9	6.3	1996	77.9	73.3	58.5	48.8	29.8	21.2	13.8	8.0
1997	72.4	68.0	53.3	43.9	25.6	17.5	11.0	6.4	1997	78.0	73.5	58.7	48.9	30.0	21.4	13.9	8.0
1998	72.6	68.2	53.5	44.2	25.8	17.8	11.1	6.5	1998	78.2	73.6	58.8	49.0	30.1	21.4	13.9	8.0
1999	72.8	68.4	53.7	44.4	26.0	18.0	11.3	6.6	1999	78.4	73.8	59.0	49.2	30.3	21.6	14.0	8.1
2000	73.1	68.6	53.9	44.6	26.3	18.2	11.5	6.6	2000	78.6	74.0	59.2	49.4	30.5	21.8	14.1	8.1
2001	73.3	68.8	54.2	44.8	26.6	18.4	11.7	6.8	2001	78.8	74.2	59.4	49.6	30.7	22.0	14.3	8.2
2002	73.5	69.0	54.3	45.0	26.7	18.6	11.8	6.8	2002	78.9	74.3	59.5	49.7	30.8	22.1	14.4	8.2
2003	73.8	69.3	54.6	45.2	27.0	18.8	12.0	6.9	2003	79.1	74.5	59.7	49.9	30.9	22.2	14.5	8.3
Northern Ireland																	
1981	69.2	65.4	50.9	41.5	23.2	15.6	9.7	5.8	1981	75.5	71.6	56.8	47.1	28.3	20.0	12.8	7.3
1986	70.9	66.8	52.2	42.7	24.2	16.4	10.4	6.2	1986	77.1	72.9	58.1	48.3	29.3	20.8	13.4	7.8
1991	72.6	68.2	53.6	44.1	25.5	17.3	11.0	6.4	1991	78.4	74.0	59.2	49.4	30.3	21.6	14.2	8.3
1996	73.8	69.4	54.7	45.3	26.6	18.2	11.4	6.6	1996	79.2	74.7	59.9	50.0	30.9	22.1	14.4	8.4
1997	74.2	69.7	55.0	45.5	26.8	18.4	11.5	6.6	1997	79.5	75.0	60.2	50.3	31.2	22.4	14.5	8.4
1998	74.3	69.8	55.2	45.7	27.0	18.6	11.6	6.6	1998	79.5	75.0	60.2	50.4	31.2	22.4	14.5	8.2
1999	74.5	70.0	55.4	45.9	27.2	18.8	11.7	6.6	1999	79.6	75.1	60.2	50.4	31.3	22.5	14.6	8.2
2000	74.8	70.4	55.7	46.2	27.6	19.1	11.9	6.6	2000	79.8	75.2	60.4	50.6	31.5	22.6	14.6	8.2
2001	75.2	70.7	56.1	46.6	27.9	19.4	12.3	6.9	2001	80.1	75.6	60.7	50.9	31.8	22.9	14.9	8.4
2002	75.6	71.1	56.4	46.9	28.2	19.7	12.4	7.0	2002	80.4	75.9	61.0	51.2	32.0	23.1	15.1	8.5
2003	75.8	71.4	56.7	47.1	28.4	19.9	12.6	7.2	2003	80.6	76.0	61.1	51.3	32.2	23.3	15.2	8.6

Note: Figures from 1981 are calculated from the population estimates revised in the light of the 2001 Census. All figures are based on a three-year period.

Table 6.1	Deaths: age and sex

England and Wales　　Numbers (thousands) and rates

Year and quarter	All ages	Under 1[1]	1–4	5–9	10–14	15–19	20–24	25–34	35–44	45–54	55–64	65–74	75–84	85 and over
Numbers (thousands)														
Males														
1976	300.1	4.88	0.88	0.68	0.64	1.66	1.66	3.24	5.93	20.4	52.0	98.7	80.3	29.0
1981	289.0	4.12	0.65	0.45	0.57	1.73	1.58	3.18	5.54	16.9	46.9	92.2	86.8	28.5
1986	287.9	3.72	0.57	0.33	0.38	1.43	1.75	3.10	5.77	14.4	43.6	84.4	96.2	32.2
1991	277.6	2.97	0.55	0.34	0.35	1.21	1.76	3.69	6.16	13.3	34.9	77.2	95.8	39.3
1996	268.7	2.27	0.44	0.24	0.29	0.93	1.41	4.06	5.84	13.6	30.1	71.0	90.7	47.8
1999	264.3	2.08	0.41	0.22	0.28	0.90	1.27	3.85	5.93	13.6	28.7	64.3	90.4	52.3
2000	255.5	1.89	0.34	0.22	0.28	0.87	1.22	3.76	6.05	13.4	27.9	60.6	87.1	51.9
2001	252.4	1.81	0.32	0.19	0.28	0.88	1.27	3.63	6.07	13.3	27.5	57.5	87.0	52.7
2002	253.1	1.81	0.32	0.20	0.28	0.83	1.24	3.47	6.20	12.9	27.7	56.3	88.3	53.6
2003	253.9	1.81	0.31	0.19	0.24	0.81	1.23	3.26	6.32	12.7	28.2	55.1	89.6	54.0
2004	244.1	1.79	0.29	0.17	0.26	0.78	1.15	3.10	6.19	12.2	27.0	52.5	87.3	51.3
2005[p]	243.9	1.88	0.30	0.17	0.27	0.86	1.22	3.15	6.36	12.2	27.3	51.1	84.7	54.5
Females														
1976	298.5	3.46	0.59	0.45	0.42	0.62	0.67	1.94	4.04	12.8	29.6	67.1	104.7	72.1
1981	288.9	2.90	0.53	0.30	0.37	0.65	0.64	1.82	3.74	10.5	27.2	62.8	103.6	73.9
1986	293.3	2.59	0.49	0.25	0.27	0.56	0.67	1.65	3.83	8.8	25.8	58.4	106.5	83.6
1991	292.5	2.19	0.44	0.25	0.22	0.46	0.64	1.73	3.70	8.4	21.3	54.2	103.3	95.7
1996	291.5	1.69	0.32	0.18	0.20	0.43	0.51	1.85	3.66	8.9	18.2	50.2	96.7	108.7
1999	291.8	1.55	0.30	0.17	0.22	0.39	0.47	1.67	3.79	9.0	18.0	45.1	93.9	117.2
2000	280.1	1.49	0.25	0.16	0.18	0.38	0.47	1.69	3.87	9.1	17.6	42.2	89.3	113.4
2001	277.9	1.43	0.27	0.19	0.18	0.38	0.47	1.59	3.77	8.9	17.6	40.5	88.8	113.9
2002	280.4	1.31	0.24	0.16	0.19	0.38	0.43	1.61	3.77	8.7	17.7	39.6	90.0	116.3
2003	284.4	1.50	0.28	0.15	0.19	0.35	0.46	1.57	3.86	8.5	18.0	39.0	92.7	117.9
2004	268.4	1.43	0.23	0.13	0.16	0.38	0.46	1.49	3.80	8.1	17.6	36.9	88.3	109.4
2005[p]	269.1	1.37	0.22	0.13	0.19	0.38	0.48	1.48	3.81	8.2	17.8	35.9	86.3	112.9
Rates (deaths per 1,000 population in each age group)														
Males														
1976	12.5	16.2	0.65	0.34	0.31	0.88	0.96	0.92	2.09	6.97	19.6	50.3	116.4	243.2
1981	12.0	12.6	0.53	0.27	0.29	0.82	0.83	0.89	1.83	6.11	17.7	45.6	105.2	226.5
1986	11.8	11.0	0.44	0.21	0.23	0.72	0.83	0.88	1.68	5.27	16.6	42.8	101.2	215.4
1991	11.2	8.3	0.40	0.21	0.23	0.72	0.89	0.94	1.76	4.56	13.9	38.1	93.1	205.6
1996	10.7	6.8	0.32	0.14	0.18	0.60	0.85	1.01	1.67	4.06	11.9	34.5	85.0	198.8
1999	10.4	6.5	0.31	0.12	0.16	0.56	0.83	0.99	1.60	3.99	10.9	31.6	79.9	194.4
2000	10.0	6.1	0.26	0.13	0.16	0.54	0.79	0.98	1.59	3.92	10.4	29.7	75.9	187.5
2001	9.9	5.9	0.25	0.11	0.16	0.53	0.80	0.97	1.56	3.89	10.0	28.0	74.0	186.4
2002	9.8	5.9	0.25	0.12	0.16	0.49	0.77	0.95	1.57	3.85	9.7	27.2	73.4	187.5
2003	9.8	5.7	0.25	0.11	0.14	0.46	0.95	0.91	1.58	3.81	9.6	26.3	72.8	190.4
2004[3]	9.4	5.5	0.23	0.10	0.15	0.44	0.68	0.88	1.53	3.67	9.0	24.9	69.8	175.2
2005[2][p]	9.3	5.7	0.24	0.10	0.16	0.48	0.70	0.89	1.56	3.61	8.9	24.0	67.2	171.6
2003 March	10.5	6.4	0.27	0.12	0.16	0.48	0.77	0.94	1.62	3.94	10.0	27.8	72.8	214.3
June	9.4	5.5	0.24	0.09	0.12	0.45	0.74	0.92	1.60	3.78	9.2	25.4	70.2	179.1
Sept	9.0	5.2	0.19	0.11	0.14	0.52	0.79	0.93	1.57	3.63	9.1	24.6	66.1	165.9
Dec	10.3	5.8	0.29	0.13	0.13	0.39	0.69	0.84	1.52	3.91	10.0	27.7	77.0	202.8
2004[3] March	10.2	5.9	0.25	0.12	0.15	0.46	0.67	0.92	1.59	3.81	9.4	26.6	76.9	199.3
June	9.1	5.2	0.23	0.12	0.14	0.39	0.74	0.94	1.58	3.72	8.8	24.5	66.9	164.0
Sept	8.7	5.3	0.23	0.10	0.18	0.46	0.71	0.86	1.47	3.58	8.5	23.2	64.5	154.8
Dec	9.5	5.5	0.23	0.08	0.11	0.43	0.58	0.78	1.49	3.58	9.2	25.4	70.8	183.0
2005[2] March[p]	10.5	6.0	0.26	0.09	0.18	0.49	0.68	0.92	1.60	3.79	9.6	26.4	77.1	202.3
June[p]	9.2	5.6	0.25	0.11	0.16	0.48	0.72	0.90	1.62	3.60	9.0	23.8	67.0	165.4
Sept[p]	8.4	5.5	0.21	0.09	0.17	0.45	0.64	0.87	1.49	3.48	8.3	22.3	59.7	146.5
Dec[p]	9.2	5.7	0.24	0.12	0.13	0.50	0.76	0.88	1.53	3.57	8.7	23.6	65.5	172.8
Females														
1976	11.8	12.2	0.46	0.24	0.21	0.35	0.40	0.56	1.46	4.30	10.1	26.0	74.6	196.6
1981	11.3	9.4	0.46	0.19	0.19	0.32	0.35	0.52	1.26	3.80	9.5	24.1	66.2	178.2
1986	11.4	8.0	0.40	0.17	0.17	0.29	0.33	0.47	1.12	3.24	9.2	23.4	62.5	169.4
1991	11.2	6.4	0.33	0.16	0.15	0.29	0.33	0.44	1.05	2.87	8.2	21.8	58.7	161.6
1996	11.0	5.3	0.25	0.10	0.12	0.29	0.31	0.46	1.04	2.63	7.1	20.6	55.8	158.9
1999	11.0	5.1	0.24	0.10	0.13	0.25	0.31	0.43	1.01	2.61	6.7	19.2	53.4	162.6
2000	10.5	5.1	0.20	0.10	0.11	0.25	0.30	0.44	1.00	2.62	6.4	18.1	50.8	155.2
2001	10.4	4.9	0.22	0.12	0.11	0.24	0.30	0.42	0.96	2.57	6.3	17.4	50.1	155.0
2002	10.4	4.5	0.20	0.10	0.11	0.24	0.27	0.44	0.94	2.54	6.0	17.0	50.4	159.4
2003	10.6	4.9	0.24	0.10	0.12	0.21	0.28	0.44	0.95	2.51	5.9	16.7	51.3	165.8
2004[3]	9.9	4.6	0.20	0.09	0.10	0.22	0.27	0.42	0.93	2.39	5.7	15.8	48.6	154.3
2005[2][p]	9.9	4.3	0.19	0.09	0.11	0.22	0.28	0.42	0.92	2.38	5.6	15.3	48.0	152.4
2003 March	11.4	5.3	0.26	0.09	0.09	0.19	0.33	0.48	1.00	2.59	6.1	17.6	54.8	184.6
June	10.0	4.8	0.24	0.09	0.17	0.22	0.25	0.43	0.90	2.58	5.8	16.1	49.3	153.6
Sept	9.6	4.5	0.20	0.12	0.10	0.21	0.30	0.43	0.97	2.38	5.6	15.3	46.8	147.6
Dec	11.2	5.2	0.26	0.09	0.10	0.24	0.25	0.40	0.94	2.49	6.2	17.8	54.3	177.5
2004[3] March	11.1	5.3	0.22	0.09	0.10	0.27	0.32	0.42	0.95	2.50	6.0	17.1	53.9	177.0
June	9.4	4.1	0.17	0.08	0.11	0.26	0.27	0.43	0.94	2.41	5.4	15.0	46.5	144.3
Sept	9.1	4.3	0.20	0.06	0.09	0.20	0.24	0.42	0.88	2.27	5.4	14.9	44.6	137.5
Dec	10.1	4.6	0.19	0.11	0.09	0.17	0.27	0.40	0.93	2.36	5.9	16.1	49.5	158.4
2005[2] March[p]	11.6	4.6	0.24	0.09	0.13	0.18	0.30	0.47	0.96	2.53	6.0	17.1	56.8	184.7
June[p]	9.7	4.7	0.17	0.11	0.10	0.25	0.29	0.39	0.97	2.35	5.5	15.3	47.4	146.6
Sept[p]	8.7	3.9	0.15	0.07	0.12	0.22	0.27	0.37	0.90	2.37	5.4	13.9	42.2	130.4
Dec[p]	9.6	4.2	0.19	0.08	0.12	0.24	0.27	0.45	0.86	2.29	5.5	15.0	45.8	148.5

Note: Figures represent the numbers of deaths registered in each year up to 1992 and the numbers of deaths occurring in each year from 1993 to 2004. Provisional figures for 2005 relate to registrations.
1　Rates per 1,000 live births.
2　Based on the 2004-based population projections for 2005.
3　Rates for 2004 have been calculated using the revised mid-2004 population estimates published on 20 December 2005.
p　provisional

Table 6.2 **Deaths: subnational**

Government Office Regions of England[1] Rates

Year and quarter		North East	North West	Yorkshire and the Humber	East Midlands	West Midlands	East	London	South East	South West
Total deaths (deaths per 1,000 population of all ages)										
1996		11.7	11.7	11.2	10.7	10.7	10.3	9.4	10.7	11.7
1997		11.6	11.6	11.1	10.5	10.6	10.2	9.0	10.6	11.7
1998		11.9	11.7	11.2	10.8	10.6	10.2	8.8	10.4	11.4
1999		11.6	11.5	10.9	10.7	10.7	10.3	8.7	10.5	11.6
2000		10.8	10.7	10.3	10.0	10.3	9.9	8.2	9.8	11.3
2001		11.1	11.0	10.4	10.1	10.2	9.9	7.9	9.9	11.0
2002		11.2	11.0	10.5	10.2	10.2	10.0	7.8	9.9	11.1
2003		11.3	11.0	10.5	10.3	10.4	9.9	7.8	9.9	11.2
2004[2]		10.9	10.5	10.1	9.7	9.8	9.5	7.2	9.4	10.4
2005[2P]		10.8	10.4	10.0	9.8	9.9	9.6	7.1	9.5	10.5
2004[2]	March	11.8	11.6	11.2	10.7	10.8	10.5	8.0	10.4	11.6
	June	10.6	10.0	9.6	9.3	9.5	9.2	7.0	9.1	9.9
	Sept	9.8	9.7	9.3	9.0	9.0	8.8	6.6	8.7	9.5
	Dec	11.2	10.6	10.3	9.9	10.1	9.6	7.4	9.6	10.7
2005[2]	March[P]	12.0	12.0	11.5	11.2	11.6	11.0	8.3	10.9	12.1
	June[P]	10.7	10.2	9.8	9.6	9.7	9.4	7.0	9.4	10.5
	Sept[P]	9.7	9.4	9.0	8.7	8.8	8.5	6.4	8.5	9.4
	Dec[P]	10.6	10.2	9.8	9.7	9.7	9.4	6.9	9.2	10.3
Infant mortality (deaths under 1 year per 1,000 live births)										
1996		6.2	6.3	6.5	6.3	6.8	5.3	6.3	5.3	5.5
1997		5.8	6.7	6.5	5.7	7.0	4.8	5.8	5.0	5.8
1998		5.0	6.3	6.9	5.6	6.5	5.0	6.0	4.4	4.8
1999		5.6	6.5	6.3	6.0	6.9	4.6	6.0	4.8	4.7
2000		6.5	6.2	7.3	5.4	6.8	4.4	5.4	4.4	4.7
2001		5.4	5.8	5.5	4.9	6.4	4.5	6.1	4.2	5.4
2002		4.8	5.4	6.1	5.6	6.6	4.3	5.5	4.5	4.3
2003		4.9	5.9	5.7	5.9	7.4	4.5	5.4	4.2	4.1
2004		4.6	5.4	5.8	4.9	6.3	4.2	5.2	3.9	4.5
2005[P]		4.5	5.8	6.1	4.7	6.4	4.0	5.1	3.9	4.5
2004	March	5.9	6.1	6.1	4.8	6.9	4.9	5.7	4.5	5.0
	June	4.6	4.9	5.8	4.8	5.6	4.0	4.6	3.3	4.8
	Sept	3.1	5.3	4.9	4.3	7.0	4.3	5.0	3.5	4.2
	Dec	4.8	5.3	6.3	5.6	5.6	3.5	5.5	4.5	4.2
2005	March[P]	4.3	5.6	6.5	6.1	6.7	4.8	5.1	4.2	4.9
	June[P]	4.5	6.4	6.8	5.5	6.1	3.8	5.8	3.5	3.7
	Sept[P]	5.2	5.0	5.5	3.6	7.1	3.8	4.9	3.7	3.9
	Dec[P]	3.8	6.3	5.7	3.6	5.8	3.7	4.5	4.3	5.7
Neonatal mortality (deaths under 4 weeks per 1,000 live births)										
1996		4.1	4.0	4.2	4.2	4.9	3.5	4.4	3.5	3.8
1997		3.7	4.3	4.4	3.7	5.0	3.3	3.7	3.4	3.9
1998		3.1	4.1	4.5	3.7	4.8	3.4	4.1	2.9	3.3
1999		4.1	4.4	4.1	4.3	4.8	3.0	4.1	3.2	3.2
2000		4.4	4.3	5.0	4.1	5.0	3.0	3.7	3.1	3.0
2001		3.5	3.8	3.2	3.4	4.4	2.9	4.1	2.9	3.7
2002		3.2	3.6	4.0	4.0	4.8	2.9	3.6	2.9	3.1
2003		3.2	4.1	4.0	4.2	5.1	3.0	3.7	2.8	2.9
2004		2.8	3.6	3.8	3.5	4.7	2.9	3.6	2.8	3.2
2005[P]		2.8	3.8	4.0	3.4	4.9	2.6	3.4	2.8	3.2
2004	March	3.7	3.5	4.0	3.5	5.3	3.4	3.9	2.7	3.8
	June	3.2	3.4	4.0	3.6	4.2	3.1	3.1	2.5	2.9
	Sept	1.4	3.8	3.2	3.3	5.5	3.0	3.5	2.6	3.0
	Dec	2.8	3.5	4.1	3.6	3.9	2.1	3.6	3.2	3.1
2005	March[P]	3.2	3.8	4.5	4.5	5.0	3.0	3.0	2.9	3.6
	June[P]	3.0	3.6	4.5	3.8	4.7	2.6	4.0	2.2	2.6
	Sept[P]	2.8	3.0	3.8	3.0	5.7	2.7	3.6	2.8	2.8
	Dec[P]	2.4	4.7	3.4	2.4	4.2	2.1	2.9	3.2	3.8
Perinatal mortality (stillbirths and deaths under 1 week per 1,000 total births)[3]										
1996		9.2	8.6	8.3	8.7	10.2	7.5	9.6	7.8	7.5
1997		8.0	8.9	8.3	7.7	9.6	7.3	9.0	7.3	8.7
1998		8.2	8.7	9.2	8.0	9.3	7.4	9.0	6.8	7.3
1999		8.2	8.7	8.3	7.8	9.9	7.0	9.0	6.9	7.8
2000		8.5	8.6	9.6	7.8	9.6	7.1	9.0	6.6	6.6
2001		7.8	8.7	7.5	7.9	9.1	7.1	8.9	6.9	7.2
2002		8.1	8.5	9.0	8.5	10.0	7.5	9.3	6.9	6.8
2003		7.8	9.0	9.0	9.5	10.2	7.3	9.5	7.0	7.0
2004[P]		7.6	8.2	8.8	8.1	9.4	7.5	8.9	7.0	7.1
2005[P]		..	..	..	..	..	..	..	..	..
2004	March	9.6	8.2	8.9	8.4	10.1	8.0	9.2	7.2	6.6
	June	8.8	8.3	9.1	8.5	8.9	7.4	8.5	6.8	7.5
	Sept	6.4	8.1	9.3	8.2	10.1	7.6	9.2	7.0	7.9
	Dec	5.7	8.4	7.8	7.2	8.3	7.0	8.6	6.9	6.6
2005	March[P]	6.5	8.0	9.5	8.8	8.8	6.7	7.6	6.2	6.7
	June[P]	8.6	7.2	9.7	7.6	10.0	6.7	7.8	6.4	6.5
	Sept[P]	7.0	6.9	7.9	7.1	10.8	5.9	8.4	6.8	5.4
	Dec[P]	..	..	..	..	..	..	..	..	..

Note: Figures represent the numbers of deaths occurring in each year with the exception of provisional figures for 2005 which relate to registrations.

Some stillbirths in 2004 are excluded from these and previously published figures, as the relevant registration details were not sent to ONS before the statistics were compiled. Revised figures for 2004 will be published as soon as possible to include the additional stillbirth registrations. See 'In brief' *Population Trends* 124.

1 The regions presented in this table have changed from the Regional Offices of the Department of Health to the Government Office Regions. See 'In brief' *Health Statistics Quarterly* 15 for details.

2 Rates for 2004 and 2005 have been calculated using the revised mid-2004 population estimates published on 20 December 2005.

3 In October 1992 the legal definition of a stillbirth was changed, from a baby born dead after 28 completed weeks of gestation or more, to one born dead after 24 completed weeks of gestation or more.

p provisional.

Table 7.1 — International migration: age and sex

United Kingdom

Numbers (thousands)

Year and quarter	All ages			0–14			15–24			25–44			45 and over		
	Persons	Males	Females	Persons	Males	Females	Persons	Males	Females	Persons	Males	Females	Persons	Males	Females
Inflow															
1971	200	103	97	33	17	17	65	28	37	81	48	33	21	10	11
1976	191	100	91	32	16	17	64	32	32	77	43	34	18	9	9
1981	153	83	71	30	16	14	48	24	24	60	34	26	15	9	7
1986	250	120	130	45	22	23	79	34	45	101	49	51	25	16	10
1991	328	157	171	53	23	30	106	47	59	139	73	66	31	14	17
1996	318	157	161	33	14	19	114	49	65	142	77	65	29	17	12
1997	326	169	157	43	22	21	126	57	68	131	76	55	27	15	12
1998	390	207	184	37	18	19	134	65	69	194	109	84	26	15	11
1999	454	250	204	42	24	18	158	79	80	224	130	94	30	18	13
2000	483	275	209	36	18	18	161	82	79	244	149	95	43	26	17
2001	480	260	219	46	25	21	158	77	81	239	135	103	37	22	14
2002	513	284	229	38	20	17	185	100	85	256	148	108	35	16	19
2003	513	261	252	41	23	18	211	99	112	219	118	101	42	21	21
2004	582	297	285	39	25	14	222	105	118	274	147	127	47	20	26
2003 March	109	51	58	9	6	3	37	15	23	54	26	28	9	4	4
June	104	49	55	10	6	4	39	14	25	45	24	21	11	4	6
Sept	205	111	94	15	8	7	101	54	47	75	40	35	15	9	6
Dec	94	50	44	8	4	4	33	15	18	45	28	18	8	3	4
2004 March	113	59	55	8	5	3	45	20	24	53	28	25	8	5	3
June	120	58	63	11	8	3	41	18	24	59	28	31	9	4	6
Sept	236	121	114	13	9	5	101	50	51	104	55	49	18	8	10
Dec	113	60	53	7	4	4	36	17	19	59	35	23	11	4	7
Outflow															
1971	240	124	116	51	26	24	64	28	36	99	57	42	27	12	15
1976	210	118	93	40	20	21	52	26	25	97	59	38	21	12	9
1981	233	133	100	49	25	24	51	29	22	108	64	44	25	14	11
1986	213	107	106	37	17	20	47	19	28	98	55	43	32	17	15
1991	285	146	139	44	19	25	76	39	37	131	69	62	33	18	15
1996	264	134	130	38	16	22	63	24	39	140	79	60	23	15	9
1997	279	153	126	29	15	13	86	45	41	138	77	61	27	16	11
1998	251	131	121	24	15	10	70	31	39	130	71	59	27	14	13
1999	291	158	133	27	19	8	87	42	45	143	79	64	34	18	16
2000	321	178	142	26	11	15	84	45	39	175	102	73	36	20	16
2001	308	173	135	25	14	11	84	41	43	155	89	65	45	29	16
2002	359	195	165	25	15	10	92	44	48	186	107	80	56	28	28
2003	362	193	169	35	19	16	85	37	47	188	105	82	55	31	24
2004	359	182	177	29	13	16	82	38	44	181	101	80	67	30	36
2003 March	76	38	39	6	3	3	17	8	9	44	22	21	10	5	5
June	77	41	36	6	3	3	19	7	12	37	23	14	15	8	7
Sept	118	63	55	18	11	7	32	13	19	53	30	23	15	9	6
Dec	90	51	39	4	3	2	17	9	8	54	30	23	15	9	6
2004 March	75	34	42	6	1	5	14	6	8	37	19	18	18	7	11
June	78	42	36	9	3	6	19	8	11	38	24	13	13	7	6
Sept	125	62	63	7	4	4	31	15	17	65	34	31	21	9	11
Dec	82	45	36	7	4	2	18	10	8	42	24	18	15	7	8
Balance															
1971	− 40	− 22	− 19	− 17	− 10	− 8	+ 1	−	+ 1	− 18	− 10	− 9	− 6	− 2	− 4
1976	− 19	− 18	− 1	− 8	− 4	− 4	+ 12	+ 6	+ 7	− 20	− 16	− 4	− 3	− 3	−
1981	− 79	− 50	− 29	− 19	− 9	− 10	− 2	− 5	+ 2	− 48	− 31	− 18	− 10	− 5	− 4
1986	+ 37	+ 13	+ 24	+ 8	+ 5	+ 3	+ 32	+ 15	+ 18	+ 3	− 5	+ 8	− 7	− 1	− 6
1991	+ 43	+ 12	+ 32	+ 8	+ 3	+ 5	+ 30	+ 9	+ 22	+ 7	+ 4	+ 4	− 2	− 4	+ 2
1996	+ 54	+ 23	+ 31	− 5	− 2	− 3	+ 51	+ 25	+ 26	+ 2	− 2	+ 5	+ 5	+ 2	+ 3
1997	+ 47	+ 16	+ 31	+ 14	+ 6	+ 8	+ 40	+ 12	+ 28	− 7	− 1	− 6	−	− 1	+ 1
1998	+ 139	+ 76	+ 63	+ 13	+ 3	+ 10	+ 64	+ 34	+ 30	+ 64	+ 38	+ 25	− 1	−	− 2
1999	+ 163	+ 92	+ 71	+ 15	+ 5	+ 10	+ 71	+ 37	+ 34	+ 81	+ 51	+ 30	− 4	− 1	− 3
2000	+ 163	+ 96	+ 66	+ 10	+ 7	+ 3	+ 77	+ 37	+ 40	+ 69	+ 47	+ 23	+ 7	+ 6	+ 1
2001	+ 172	+ 88	+ 84	+ 21	+ 11	+ 10	+ 74	+ 36	+ 38	+ 84	+ 46	+ 38	− 8	− 6	− 2
2002	+ 153	+ 89	+ 64	+ 13	+ 5	+ 8	+ 93	+ 56	+ 37	+ 69	+ 41	+ 28	−22	− 13	− 9
2003	+ 151	+ 68	+ 83	+ 7	+ 4	+ 2	+ 126	+ 61	+ 65	+ 31	+ 12	+ 19	− 13	− 10	− 3
2004	+ 223	+ 115	+ 108	+ 10	+ 12	− 2	+ 140	+ 67	+ 73	+ 93	+ 46	+ 47	− 20	− 10	− 10
2003 March	+ 33	+ 13	+ 20	+ 3	+ 3	−	+ 20	+ 6	+ 14	+ 11	+ 4	+ 6	− 1	− 1	− 1
June	+ 27	+ 8	+ 19	+ 4	+ 3	+ 1	+ 20	+ 7	+ 13	+ 7	+ 1	+ 6	− 4	− 4	− 1
Sept	+ 87	+ 48	+ 39	− 4	− 3	− 1	+ 69	+ 42	+ 28	+ 21	+ 10	+ 12	−	−	−
Dec	+ 4	−1	+ 5	+ 3	+ 1	+ 2	+ 16	+ 6	+ 10	− 8	− 3	− 6	− 7	− 6	− 1
2004 March	+ 38	+ 25	+ 13	+ 2	+ 4	− 2	+ 30	+ 14	+ 16	+ 16	+ 9	+ 7	− 10	− 2	− 8
June	+ 42	+ 16	+ 26	+ 2	+ 5	− 3	+ 22	+ 10	+ 12	+ 22	+ 4	+ 17	− 3	− 3	+ 0
Sept	+ 111	+ 60	+ 51	+ 6	+ 5	+ 1	+ 69	+ 36	+ 34	+ 39	+ 21	+ 17	− 3	− 2	− 1
Dec	+ 31	+ 14	+ 17	+ 1	− 1	+ 1	+ 18	+ 7	+ 11	+ 16	+ 11	+ 5	− 4	− 3	− 1

Note: Figures in this table are derived from the International Passenger Survey and other sources – see Notes to Tables. Prior to 1991 they exclude certain categories of migration such as migrants between the UK and the Irish Republic, persons seeking asylum after entering the country and other short-term visitors granted extensions of stay. From 1991, the figures in this table include all categories of migrants and therefore represent Total International Migration. For adjustments required to pre-1991 figures, see Notes to Tables.

Table 7.2 International migration: country of last or next residence

United Kingdom Numbers (thousands)

Year and quarter	All countries	European Union[1]	Commonwealth countries						Other foreign countries		
			Australia, New Zealand, Canada	South Africa	India, Bangladesh, Sri Lanka[2]	Pakistan[2]	Caribbean	Other	USA	Middle East[3]	Other[3]
Inflow											
1971	200	21	52	8	24	:	5	36	22	–	31
1976	191	33	40	9	15	12	4	32	16	7	23
1981	153	25	20	3	18	9	3	19	17	11	27
1986	250	72	30	18	16	10	5	25	26	15	34
1991	328	95	44	8	17	16	4	42	24	11	69
1996	318	98	37	11	15	11	4	33	32	13	63
1997	326	100	40	13	21	9	4	32	23	15	67
1998	390	109	64	20	17	10	6	31	37	13	84
1999	454	99	63	29	25	12	6	37	29	15	138
2000	483	96	63	23	34	16	6	48	24	30	144
2001	480	86	77	22	32	18	3	47	24	30	140
2002	513	89	61	27	36	10	5	52	28	32	172
2003	513	101	68	28	45	13	4	49	28	27	150
2004	582	139	63	37	60	29	6	60	28	26	135
2003 March	109	26	15	8	8	1	1	9	5	7	31
June	104	16	17	7	11	4	1	8	6	5	28
Sept	205	41	23	7	17	4	2	23	15	9	65
Dec	94	18	14	6	9	3	–	10	3	5	26
2004 March	113	23	14	10	12	10	2	14	3	5	20
June	120	30	19	9	14	4	1	9	5	4	25
Sept	236	52	19	11	21	10	2	26	16	12	67
Dec	113	33	10	7	13	5	0	11	4	6	24
Outflow											
1971	240	31	99	21	8	:	8	23	17	:	34
1976	210	39	63	21	4	2	3	17	21	6	33
1981	232	33	78	23	2	1	3	20	25	23	23
1986	213	62	50	2	4	2	2	13	34	16	28
1991	285	95	61	7	6	4	2	21	35	14	40
1996	264	94	58	5	5	1	1	23	26	8	42
1997	279	92	57	8	6	3	3	23	28	13	46
1998	251	85	54	6	5	2	2	14	27	9	48
1999	291	103	73	7	4	1	3	14	33	10	44
2000	321	103	79	7	5	3	3	15	33	15	58
2001	308	94	80	8	8	3	2	13	28	9	63
2002	359	125	84	10	7	4	2	16	37	12	62
2003	362	122	90	14	7	4	1	15	27	7	75
2004	359	121	95	10	6	4	3	20	27	12	61
2003 March	76	30	20	2	1	1	–	2	4	1	16
June	77	32	18	1	2	1	–	1	5	1	15
Sept	118	41	25	5	2	1	–	6	9	3	25
Dec	90	19	27	5	2	1	–	5	10	1	19
2004 March	75	26	21	2	1	0	0	3	7	3	11
June	78	32	19	1	1	2	1	3	6	2	10
Sept	125	45	29	1	2	1	2	8	8	5	25
Dec	82	19	26	5	2	1	1	6	6	2	15
Balance											
1971	– 40	– 10	– 46	– 13	+ 16	:	– 3	+ 14	+ 6	:	– 3
1976	– 19	– 6	– 23	– 12	+ 12	+ 10	–	+ 15	– 4	+ 1	– 10
1981	– 79	– 8	– 58	– 20	+ 15	+ 8	+ 1	– 2	– 8	– 12	+ 5
1986	+ 37	+ 9	– 21	+ 16	+ 12	+ 8	+ 3	+ 12	– 8	–	+ 6
1991	+ 43	–	– 18	+ 1	+ 11	+ 12	+ 2	+ 20	– 11	– 4	+ 29
1996	+ 54	+ 5	– 21	+ 6	+ 10	+ 10	+ 3	+ 10	+ 6	+ 5	+ 21
1997	+ 47	+ 9	– 17	+ 5	+ 15	+ 6	+ 1	+ 9	– 5	+ 2	+ 21
1998	+ 139	+ 24	+ 10	+ 14	+ 12	+ 8	+ 4	+ 17	+ 10	+ 4	+ 36
1999	+ 163	– 4	– 10	+ 22	+ 22	+ 11	+ 3	+ 23	– 4	+ 5	+ 94
2000	+ 163	– 8	– 15	+ 15	+ 29	+ 13	+ 4	+ 33	– 9	+ 15	+ 86
2001	+ 172	– 7	– 2	+ 13	+ 24	+ 14	+ 1	+ 34	– 4	+ 20	+ 77
2002	+ 153	– 36	– 23	+ 17	+ 29	+ 7	+ 3	+ 36	– 10	+ 20	+ 110
2003	+ 151	– 21	– 22	+ 14	+ 38	+ 9	+ 3	+ 34	+ 1	+ 20	+ 75
2004	+ 223	+ 17	– 32	+ 27	+ 54	+ 25	+ 2	+ 40	–	+ 14	+ 74
2003 March	+ 33	– 4	– 5	+ 5	+ 7	+ 1	+ 1	+ 6	+ 1	+ 6	+ 15
June	+ 27	– 15	– 2	+ 6	+ 10	+ 3	+ 1	+ 6	+ 1	+ 4	+ 14
Sept	+ 87	–	– 2	+ 2	+ 15	+ 2	+ 2	+ 17	+ 6	+ 5	+ 40
Dec	+ 4	– 1	– 14	+ 1	+ 7	+ 3	–	+ 5	– 7	+ 4	+ 7
2004 March	+ 38	– 3	– 7	+ 8	+ 11	+ 10	+ 2	+ 11	– 4	+ 2	+ 9
June	+ 42	– 2	+ 0	+ 8	+ 13	+ 2	+ 0	+ 6	– 1	+ 2	+ 15
Sept	+ 111	+ 8	– 10	+ 10	+ 19	+ 9	+ 0	+ 18	+ 8	+ 7	+ 42
Dec	+ 31	+ 15	– 16	+ 1	+ 11	+ 4	+ 0	+ 6	– 2	+ 4	+ 9

Note: Figures in this table are derived from the International Passenger Survey and other sources – see Notes to Tables. Prior to 1991 they exclude certain categories of migration such as migrants between the UK and the Irish Republic, persons seeking asylum after entering the country and other short-term visitors granted extensions of stay. From 1991, the figures in this table include all categories of migrants and therefore represent Total International Migration. For adjustments required to pre-1991 figures, see Notes to Tables.

1 For 1971 the European Union figures are for the original six countries only. From 1976 up to and including 2003, estimates are shown for the EU15 (Austria, Belgium, Denmark, Finland, France, Germany, Greece, the Irish Republic, Italy, Luxembourg, Netherlands, Portugal, Spain and Sweden). For 2004, the estimates are for the EU25 (EU15 plus the 10 countries of Cyprus, the Czech Republic, Estonia, Hungary, Latvia, Lithuania, Malta, Poland, Slovakia and Slovenia). These countries are included in the definition for the whole of 2004, whether migration occurred before or after 1 May.
2 For 1971 Pakistan is included with India, Bangladesh and Sri Lanka.
3. For 2004, the Other Commonwealth excludes Malta and Cyprus.
4 For 1971 Middle East is included in the 'Other' category of 'Other foreign' countries.

Table 7.3 | **International migration: citizenship**

United Kingdom Numbers (thousands)

| Year and quarter | Citizenship (numbers) | | | | | | | | British citizens as percentage of all citizens |
| | All countries | British | Non-British | European Union[1] | Commonwealth | | | Other foreign[3] | |
					All	Old	New[2]		
Inflow									
1971	200	92	108	..	53	17	36	54	46
1976	191	87	104	19	57	17	40	28	45
1981	153	60	93	12	43	12	31	38	39
1986	250	120	130	36	50	19	31	44	48
1991	328	109	219	53	85	26	59	82	33
1996	318	94	224	72	78	29	49	73	29
1997	326	89	237	72	90	31	59	76	27
1998	390	103	287	82	105	54	51	100	26
1999	454	116	337	67	121	54	66	150	26
2000	483	104	379	63	148	57	91	168	22
2001	480	106	373	60	151	67	84	162	22
2002	513	95	418	63	159	66	93	197	18
2003	513	106	407	64	166	63	103	177	21
2004	582	88	494	117	219	76	143	158	15
2003 March	109	26	83	14	31	15	17	38	24
June	104	22	83	14	37	16	21	32	21
Sept	205	39	166	27	62	21	40	77	19
Dec	94	19	75	10	36	11	25	30	20
2004 March	113	20	92	17	51	18	33	25	18
June	120	19	100	26	49	23	26	26	16
Sept	236	33	200	48	76	22	54	78	14
Dec	113	15	97	26	43	14	30	28	13
Outflow									
1971	240	171	69	..	29	13	16	40	71
1976	210	137	73	18	30	16	13	25	65
1981	232	164	68	16	29	14	15	24	71
1986	213	132	81	13	29	19	10	40	62
1991	285	154	131	53	35	18	17	43	54
1996	264	156	108	44	32	17	14	32	59
1997	279	149	131	53	40	20	20	38	53
1998	251	126	126	49	33	20	13	44	50
1999	291	139	152	59	41	29	12	52	48
2000	321	161	160	57	47	32	15	55	50
2001	308	159	149	49	51	32	19	49	52
2002	359	186	174	52	58	42	16	64	52
2003	362	191	171	50	59	42	17	62	53
2004	359	208	152	43	55	35	20	54	58
2003 March	76	40	36	16	8	5	3	13	53
June	77	40	37	15	11	9	2	11	51
Sept	118	70	48	11	16	10	6	21	59
Dec	90	41	49	8	23	17	6	18	45
2004 March	75	50	24	6	8	6	2	11	67
June	78	40	37	15	10	6	4	12	51
Sept	125	78	47	15	15	9	7	17	62
Dec	82	39	42	7	21	14	7	14	48
Balance									
1971	− 40	− 79	+ 39	..	+ 24	+ 4	+ 20	+ 14	:
1976	− 19	− 50	+ 31	+ 1	+ 27	+ 1	+ 27	+ 3	:
1981	− 79	−104	+ 24	− 4	+ 14	− 2	+ 16	+ 15	:
1986	+ 37	− 11	+ 49	+ 22	+ 21	+ 0	+ 21	+ 5	:
1991	+ 43	− 45	+ 89	+ 0	+ 50	+ 7	+ 42	+ 39	:
1996	+ 54	− 62	+ 116	+ 28	+ 47	+12	+ 35	+ 41	:
1997	+ 47	− 60	+ 107	+ 18	+ 50	+11	+ 39	+ 38	:
1998	+139	− 23	+ 162	+ 33	+ 72	+34	+ 38	+ 57	:
1999	+163	− 23	+ 186	+ 8	+ 80	+26	+ 54	+ 98	:
2000	+163	− 57	+ 220	+ 6	+101	+25	+ 76	+113	:
2001	+172	− 53	+ 225	+ 11	+101	+35	+ 65	+113	:
2002	+153	− 91	+ 245	+ 11	+101	+23	+ 77	+133	:
2003	+151	− 85	+ 236	+ 14	+107	+21	+ 86	+115	:
2004	+223	−120	+ 342	+ 74	+164	+41	+123	+104	:
2003 March	+ 33	− 14	+ 47	− 2	+ 23	+ 9	+ 14	+ 25	:
June	+ 27	− 18	+ 45	− 2	+ 26	+ 7	+ 19	+ 21	:
Sept	+ 87	− 31	+ 118	+ 16	+ 45	+11	+ 34	+ 56	:
Dec	+ 4	− 22	+ 26	+ 1	+ 13	− 6	+ 19	+ 12	:
2004 March	+ 38	− 30	+ 67	+ 11	+ 42	+11	+ 31	+ 15	:
June	+ 42	− 21	+ 62	+ 11	+ 38	+16	+ 22	+ 14	:
Sept	+111	− 44	+ 153	+ 33	+ 61	+14	+ 47	+ 61	:
Dec	+ 31	− 24	+ 55	+ 19	+ 22	+ 0	+ 22	+ 14	:

Note: Figures in this table are derived from the International Passenger Survey and other sources – see Notes to Tables. Prior to 1991 they exclude certain categories of migration such as migrants between the UK and the Irish Republic, persons seeking asylum after entering the country and other short-term visitors granted extensions of stay. From 1991, the figures in this table include all categories of migrants and therefore represent Total International Migration. For adjustments required to pre-1991 figures, see Notes to Tables.

1 For 1971 citizens of the European Union are included in 'Other foreign' category. From 1976 up to and including 2003, estimates are shown for the EU15 (Austria, Belgium, Denmark, Finland, France, Germany, Greece, the Irish Republic, Italy, Luxembourg, Netherlands, Portugal, Spain and Sweden). For 2004, the estimates are for the EU25 (EU15 plus the 10 countries of Cyprus, the Czech Republic, Estonia, Hungary, Latvia, Lithuania, Malta, Poland, Slovakia and Slovenia). These countries are included in the definition for the whole of 2004, whether migration occurred before or after 1 May.
2 For 2004, the New Commonwealth excludes Malta and Cyprus.
3 For 2004 onwards, Other foreign excludes the eight central and eastern European member states that joined the EU in May 2004.

Table 8.1 — Internal migration

Recorded movements between constituent countries of the United Kingdom and Government Office Regions of England

Numbers (thousands)

Year and quarter	England	Wales	Scotland	Northern Ireland	Government Office Regions of England								
					North East	North West	Yorkshire and The Humber	East Midlands	West Midlands	East	London	South East	South West
Inflow													
1976	105.4	52.0	50.4	9.7	39.2	93.0	78.2	84.0	75.7	146.3	..	215.4	123.8
1981	93.7	44.6	45.4	6.8	31.1	79.3	68.3	76.6	66.9	121.4	155.0	201.8	108.3
1986	115.6	55.2	43.9	8.8	36.5	90.0	78.6	101.9	87.1	144.6	182.8	243.3	148.8
1991	95.8	51.5	55.8	12.5	40.2	96.1	85.0	89.6	82.7	122.1	148.8	197.6	120.7
1994	103.4	52.0	51.7	10.9	37.1	99.7	87.6	96.4	84.8	130.6	160.4	215.5	127.7
1995	108.1	54.7	48.5	14.1	37.9	103.7	90.8	101.3	90.0	134.6	170.7	218.6	131.6
1996	111.1	55.3	47.0	11.4	38.6	105.0	90.8	102.1	90.6	139.5	168.0	228.0	138.5
1997	110.9	58.5	55.3	10.2	38.6	106.5	92.6	107.7	92.7	145.0	167.3	229.6	144.0
1998	111.2	56.3	52.6	11.7	39.0	104.0	93.0	107.9	93.4	142.8	173.9	226.1	138.7
1999	111.7	58.0	50.9	11.6	38.7	105.4	95.2	111.3	93.7	148.4	162.9	228.6	143.2
2000	108.6	59.5	48.8	11.2	39.2	106.2	96.5	112.1	94.3	145.8	163.0	224.2	140.1
2001	104.2	60.0	56.5	12.7	40.4	106.3	96.5	115.5	95.3	147.2	159.7	223.8	143.3
2002	100.9	64.0	52.7	10.8	42.7	108.9	99.7	119.5	98.6	150.0	154.8	228.6	145.9
2003	97.5	62.7	59.8	12.1	41.9	109.3	99.4	114.8	95.0	144.6	148.3	220.5	141.6
2004	96.6	60.1	56.8	12.5	40.7	104.9	98.1	111.8	95.1	145.5	155.1	223.4	138.8
2004 March	20.7	12.7	12.9	3.0	8.0	22.1	19.0	22.6	19.9	32.6	34.2	47.1	29.0
June	22.2	13.4	14.7	3.2	8.2	23.5	20.1	24.3	21.4	34.9	36.0	51.3	31.5
Sept	31.9	21.5	15.1	3.4	15.9	36.2	39.3	41.8	32.1	46.0	48.2	75.4	47.1
Dec	21.9	12.5	14.1	3.0	8.5	23.1	19.7	23.0	21.7	32.1	36.7	49.6	31.3
2005 March	21.0	11.3	12.7	3.4	7.6	21.5	18.4	20.5	19.4	29.4	36.2	43.9	26.1
June	22.0	11.7	15.4	3.3	7.8	21.8	18.4	22.0	20.9	31.8	36.5	47.5	28.8
Outflow													
1976	104.8	43.9	54.5	14.2	40.2	102.9	78.5	77.2	89.5	115.6	..	181.7	94.7
1981	91.5	41.8	47.7	9.4	39.1	98.6	73.3	71.7	78.4	104.4	187.0	166.0	88.0
1986	100.7	49.8	57.9	15.1	45.6	115.8	90.5	84.8	94.8	128.1	232.4	204.1	102.5
1991	112.2	47.4	46.7	9.3	40.9	104.9	85.4	81.4	87.9	113.0	202.1	184.6	98.9
1994	106.3	50.4	49.0	12.2	43.5	109.8	91.9	86.2	95.1	115.5	206.3	190.4	103.9
1995	107.9	53.1	52.0	12.3	45.6	115.8	97.6	91.9	98.1	118.7	207.6	195.8	108.0
1996	105.3	53.3	54.5	11.8	44.5	114.0	98.2	94.3	101.0	121.1	213.4	198.9	109.8
1997	114.8	54.4	53.2	12.6	44.5	117.5	100.0	97.4	103.7	124.8	221.7	205.7	112.4
1998	111.3	54.2	53.8	12.4	43.7	115.8	97.9	97.3	100.9	125.0	217.9	209.4	110.9
1999	111.6	53.3	54.9	12.5	43.8	114.9	97.0	96.4	101.8	125.8	228.3	208.7	110.7
2000	110.8	52.1	53.3	11.9	42.9	111.3	95.7	94.9	101.5	124.6	231.5	210.5	110.7
2001	120.4	51.5	50.4	11.1	42.6	110.4	95.6	95.6	101.6	127.1	244.2	216.4	110.7
2002	119.3	49.7	48.4	11.1	41.3	107.5	94.6	96.9	102.7	130.1	262.5	220.2	111.0
2003	126.0	48.1	46.4	11.7	40.1	104.1	93.0	96.0	101.7	127.4	262.6	211.1	108.0
2004	121.5	49.2	45.1	10.2	39.4	104.1	92.2	97.0	100.7	128.3	260.2	208.1	108.4
2004 March	26.6	10.1	10.2	2.3	7.8	21.2	18.9	20.1	20.6	26.7	58.8	43.9	22.6
June	29.6	11.3	10.4	2.1	9.2	24.3	21.6	22.7	22.8	28.2	59.8	46.3	23.8
Sept	37.7	16.7	14.0	3.4	13.6	35.8	31.5	33.1	35.5	45.0	82.8	72.2	38.5
Dec	27.5	11.1	10.5	2.3	8.8	22.8	20.2	21.1	21.8	28.4	58.9	45.7	23.5
2005 March	25.4	10.3	10.0	2.8	8.2	20.8	18.4	19.5	19.6	25.4	52.7	41.5	21.3
June	28.0	11.4	10.4	2.7	9.0	23.2	21.8	22.0	21.5	25.6	52.6	43.0	22.7
Balance													
1976	+ 0.6	+ 8.1	− 4.1	− 4.5	− 1.0	− 9.8	− 0.3	+ 6.8	−13.8	+30.7	..	+ 33.7	+29.1
1981	+ 2.1	+ 2.7	− 2.3	− 2.5	− 8.0	−19.3	− 5.0	+ 4.9	−11.6	+17.0	− 32.0	+ 35.8	+20.3
1986	+14.9	+ 5.4	−14.1	− 6.3	− 9.1	−25.8	−11.9	+17.1	− 7.8	+16.5	− 49.6	+ 39.2	+46.4
1991	−16.4	+ 4.0	+ 9.2	+ 3.2	− 0.7	− 8.8	− 0.4	+ 8.1	− 5.2	+ 9.1	− 53.3	+ 13.0	+21.8
1994	− 2.9	+ 1.5	+ 2.6	− 1.2	− 6.4	−10.1	− 4.4	+10.2	−10.3	+15.1	− 45.9	+ 25.1	+23.8
1995	+ 0.2	+ 1.6	− 3.5	+ 1.8	− 7.7	−12.1	− 6.8	+ 9.4	− 8.1	+15.9	− 36.9	+ 22.7	+23.6
1996	+ 5.8	+ 2.0	− 7.5	− 0.4	− 5.9	− 9.0	− 7.4	+ 7.8	−10.4	+18.3	− 45.4	+ 29.1	+28.7
1997	− 3.8	+ 4.1	+ 2.2	− 2.4	− 5.9	−11.0	− 7.3	+10.3	−11.1	+20.3	− 54.4	+ 23.8	+31.6
1998	− 0.1	+ 2.1	− 1.2	− 0.8	− 4.8	−11.8	− 4.9	+10.6	− 7.4	+17.7	− 44.0	+ 16.7	+27.8
1999	+ 0.1	+ 4.7	− 4.0	− 0.8	− 5.1	− 9.5	− 1.8	+14.9	− 8.1	+22.6	− 65.4	+ 19.8	+32.6
2000	− 2.2	+ 7.4	− 4.5	− 0.7	− 3.7	− 5.1	+ 0.8	+17.2	− 7.2	+21.2	− 68.6	+ 13.8	+29.3
2001	−16.3	+ 8.5	+ 6.1	+ 1.6	− 2.3	− 4.1	+ 0.9	+19.9	− 6.3	+20.1	− 84.5	+ 7.4	+32.6
2002	−18.4	+14.3	+ 4.3	− 0.3	+ 1.4	+ 1.4	+ 5.0	+22.6	− 4.1	+19.9	−107.8	+ 8.4	+34.8
2003	−28.5	+14.6	+13.4	+ 0.4	+ 1.8	+ 5.2	+ 6.4	+18.7	− 6.7	+17.2	−114.3	+ 9.4	+33.6
2004	−25.0	+10.9	+11.7	+ 2.3	+ 1.3	+ 0.8	+ 5.9	+14.8	− 5.6	+17.2	−105.1	+ 15.3	+30.5
2004 March	− 6.0	+ 2.6	+ 2.7	+ 0.7	+ 0.2	+ 0.9	+ 0.2	+ 2.5	− 0.7	+ 5.9	− 24.6	+ 3.3	+ 6.4
June	− 7.4	+ 2.1	+ 4.3	+ 1.1	− 1.0	− 0.8	− 1.5	+ 1.6	− 1.4	+ 6.7	− 23.7	+ 5.0	+ 7.7
Sept	− 5.9	+ 4.8	+ 1.1	− 0.0	+ 2.3	+ 0.4	+ 7.9	+ 8.7	− 3.4	+ 1.0	− 34.6	+ 3.1	+ 8.6
Dec	− 5.7	+ 1.4	+ 3.6	+ 0.6	− 0.3	+ 0.3	− 0.5	+ 2.0	− 0.2	+ 3.6	− 22.2	+ 3.9	+ 7.8
2005 March	− 4.4	+ 1.0	+ 2.8	+ 0.7	− 0.6	+ 0.6	− 0.0	+ 1.0	− 0.1	+ 4.0	− 16.5	+ 2.4	+ 4.8
June	− 5.9	+ 0.3	+ 5.0	+ 0.7	− 1.2	− 1.5	− 3.5	− 0.0	− 0.6	+ 6.2	− 16.1	+ 4.5	+ 6.1

Notes: Figures are derived from re-registrations recorded at the National Health Service Central Register.
See Notes to tables for effects of computerisation of National Health Service Central Register at Southport on time series data.
Figures have been adjusted for minor changes caused by database realignment during HA reorganisation. See Notes to tables.

| Table 9.1 | First marriages[1]: age and sex |

England and Wales — Numbers (thousands), rates, percentages, mean and median age

Year and quarter	All ages		Persons marrying per 1,000 single population at ages						Per cent aged under 20	Mean age[3] (years)	Median age[3] (years)
	Number	Rate[2]	16–19	20–24	25–29	30–34	35–44	45 and over			
Males											
1961	308.8	74.9	16.6	159.1	182.8	91.9	39.8	9.3	6.9	25.6	24.0
1966	339.1	78.9	22.1	168.6	185.4	91.1	36.4	8.6	9.9	24.9	23.4
1971	343.6	82.3	26.1	167.7	167.3	84.6	33.8	8.0	10.1	24.6	23.4
1976	274.4	62.8	18.5	123.7	132.5	78.7	32.0	7.1	9.8	25.1	23.7
1981	259.1	51.7	11.1	94.1	120.8	70.3	31.1	5.4	7.2	25.4	24.1
1986	253.0	45.0	6.0	64.4	105.1	73.9	30.9	4.8	3.8	26.3	25.1
1991	222.8	37.8	3.4	43.3	81.0	66.5	29.9	4.8	2.1	27.5	26.5
1994	206.1	34.3	2.2	31.7	73.3	61.1	30.2	5.1	1.3	28.5	27.5
1995	198.2	32.4	2.0	28.3	68.2	59.9	30.2	5.0	1.2	28.9	27.9
1996	193.3	31.1	1.8	25.2	64.5	59.4	30.7	5.2	1.1	29.3	28.3
1997	188.3	29.7	1.8	22.8	61.1	58.0	30.6	5.2	1.2	29.6	28.6
1998	186.3	28.9	1.7	21.0	59.4	57.8	30.2	5.2	1.2	29.8	28.9
1999	184.3	28.0	1.7	18.9	56.9	57.7	30.4	5.3	1.2	30.1	29.2
2000	186.1	27.7	1.7	18.2	54.3	58.2	32.0	5.7	1.2	30.5	29.6
2001	175.7	25.5	1.5	16.2	50.4	54.5	29.6	5.3	1.1	30.6	29.7
2002	179.1	25.3	1.3	16.4	48.9	55.0	31.1	5.9	1.0	30.9	30.1
2003	189.5	26.1	1.3	16.3	49.8	57.6	32.7	6.9	1.0	31.2	30.3
2004[p]	190.1	25.5	1.2	15.7	47.9	57.0	33.0	7.1	0.9	31.4	30.4
2002 March	20.7	11.9	1.1	8.8	21.3	24.1	15.1	3.5	1.7	31.0	30.0
June	49.7	28.3	1.3	17.4	54.9	61.7	34.9	6.5	0.9	31.0	30.1
Sept	77.8	43.8	1.8	27.9	88.3	95.5	50.8	8.0	0.8	30.7	29.9
Dec	31.0	17.4	1.1	11.3	30.5	37.9	23.3	5.3	1.3	31.4	30.4
2003 March	22.3	12.5	1.1	8.8	21.8	25.7	16.7	4.3	1.7	31.4	30.4
June	52.3	28.9	1.4	17.5	55.5	64.1	36.4	7.5	0.9	31.2	30.3
Sept	82.1	44.8	1.7	27.5	89.8	100.7	52.9	9.8	0.7	31.0	30.1
Dec	32.8	17.9	1.1	11.1	31.5	39.2	24.6	6.0	1.2	31.6	30.7
2004 March[p]	23.4	12.6	1.0	9.2	22.2	26.3	16.7	4.1	1.6	31.4	30.3
June[p]	51.9	28.0	1.1	16.7	52.3	63.3	36.9	8.0	0.8	31.5	30.6
Sept[p]	82.2	43.8	1.6	26.0	86.7	99.8	54.0	10.4	0.7	31.2	30.3
Dec[p]	32.6	17.4	1.1	10.8	30.3	38.6	24.2	5.8	1.2	31.7	30.7
Females											
1961	312.3	83.0	77.0	261.1	162.8	74.6	29.8	4.6	28.7	23.1	21.6
1966	342.7	89.3	82.6	263.7	153.4	74.1	30.2	4.3	32.5	22.5	21.2
1971	347.4	97.0	92.9	246.5	167.0	75.7	30.3	4.8	31.1	22.6	21.4
1976	276.5	76.9	66.7	185.4	140.7	77.6	31.6	4.0	31.1	22.8	21.5
1981	263.4	64.0	41.5	140.8	120.2	67.0	28.7	2.8	24.1	23.1	21.9
1986	256.8	55.6	24.1	102.4	108.7	67.1	28.6	2.7	13.9	24.1	23.1
1991	224.8	46.7	14.0	73.0	90.6	62.7	28.1	2.8	7.9	25.5	24.6
1994	206.3	41.6	9.6	56.4	84.5	58.9	27.7	3.1	5.2	26.5	25.7
1995	198.6	39.3	9.0	50.8	80.5	57.1	27.6	3.1	5.1	26.8	26.0
1996	192.7	37.3	8.0	45.7	77.2	57.2	27.8	3.2	4.9	27.2	26.4
1997	188.5	35.6	7.4	42.5	74.1	56.1	27.2	3.3	4.7	27.5	26.7
1998	187.4	34.7	7.2	39.9	72.6	56.1	26.5	3.4	4.7	27.7	27.0
1999	185.3	33.5	6.7	36.7	70.8	56.0	26.5	3.5	4.4	28.0	27.3
2000	187.7	33.2	6.5	35.2	68.7	57.2	27.5	3.9	4.2	28.2	27.5
2001	177.5	30.6	5.5	31.9	64.3	53.2	25.5	3.7	3.9	28.4	27.7
2002	180.7	30.3	5.3	31.0	63.2	54.4	26.8	4.3	3.7	28.7	27.9
2003	191.2	31.2	5.3	31.3	64.4	57.3	28.4	5.2	3.6	28.9	28.1
2004[p]	192.4	30.5	4.9	29.7	62.8	57.6	28.5	5.5	3.4	29.1	28.3
2002 March	20.6	14.1	4.0	14.8	26.1	24.1	13.7	2.8	6.0	28.7	27.9
June	50.1	33.8	5.3	33.7	71.3	61.0	30.2	4.7	3.3	28.8	28.0
Sept	78.8	52.6	7.4	55.5	115.1	92.3	41.4	5.8	3.0	28.5	27.7
Dec	31.1	20.7	4.7	19.5	39.4	39.5	21.7	4.0	4.8	29.2	28.4
2003 March	22.1	14.7	4.2	15.8	25.5	25.2	15.5	3.6	6.1	29.0	28.0
June	53.0	34.7	5.6	33.8	73.3	63.7	31.3	5.8	3.4	29.0	28.2
Sept	83.3	54.0	6.9	55.1	118.1	98.4	44.7	6.9	2.7	28.7	28.0
Dec	32.7	21.2	4.6	20.0	39.8	41.4	21.7	4.4	4.6	29.3	28.6
2004 March[p]	23.2	14.8	4.5	15.8	26.2	25.6	14.8	3.6	6.4	28.9	28.0
June[p]	52.5	33.5	4.8	31.7	69.4	64.4	33.9	6.4	3.0	29.3	28.4
Sept[p]	83.7	52.8	6.2	52.4	115.9	98.5	45.0	7.4	2.5	28.9	28.1
Dec[p]	33.1	20.9	4.0	18.8	39.6	41.8	22.4	4.8	4.0	29.6	28.8

Notes: Marriage rates for 1986 have been calculated using the interim revised marital status estimates (based on the original mid-2001 population estimates) and are subject to further revision.

1 Figures for all marriages can be found in Table 2.1.
2 Per 1,000 single persons aged 16 and over.
3 The mean/median ages shown in this table are unstandardised and therefore take no account of changes in the structure of the population by age or marital status.
p provisional

See 'Notes to tables'.

Table 9.2 **Remarriages[1]: age, sex, and previous marital status**

England and Wales

Numbers (thousands), rates, percentages, mean and median age

Year and quarter		Remarriages of divorced persons										Remarriages of widowed persons	
		All ages		Persons remarrying per 1,000 divorced population at ages					Per cent aged under 35	Mean[3] age (years)	Median[3] age (years)		
		Number	Rate[2]	16–24	25–29	30–34	35–44	45 and over				Number	Rate[4]
Males													
1961		18.8	162.9	478.6	473.6	351.6	198.3	88.6	33.9	40.5	39.2	19.1	28.8
1966		26.7	192.2	737.8	522.5	403.1	244.4	89.4	40.8	39.3	37.4	18.7	28.3
1971		42.4	227.3	525.2	509.0	390.7	251.3	124.8	42.8	39.8	37.0	18.7	27.5
1976		67.2	178.8	656.8	359.7	266.8	187.9	94.0	46.7	38.4	36.0	16.9	24.7
1981		79.1	129.5	240.7	260.9	205.8	141.9	63.9	46.1	38.1	35.9	13.8	19.7
1986		83.4	91.0	141.4	158.9	141.3	106.0	49.9	38.5	39.1	37.7	11.6	16.7
1991		74.9	63.0	81.1	111.3	100.6	72.7	38.4	34.3	40.3	39.0	9.0	12.5
1994		76.6	60.0	180.6	131.7	110.2	71.5	36.1	31.5	41.1	39.6	8.4	11.5
1995		77.0	58.6	190.0	132.1	111.4	72.2	34.9	30.3	41.3	39.8	7.8	10.8
1996		78.0	57.9	166.2	135.2	111.2	73.8	35.0	28.2	41.7	40.2	7.7	10.6
1997		76.8	55.7	170.9	132.2	110.3	72.9	33.6	27.0	42.0	40.5	7.4	10.3
1998		74.0	52.7	167.0	124.7	104.1	71.6	32.0	24.8	42.4	40.8	6.9	9.6
1999		72.6	50.7	125.7	120.7	102.9	70.2	31.2	23.3	42.7	41.2	6.6	9.3
2000		75.4	51.8	97.9	113.2	103.6	74.4	32.6	20.8	43.2	41.8	6.5	9.1
2001		67.7	45.7	75.7	96.6	95.8	67.6	28.5	19.7	43.5	42.0	5.8	8.0
2002		70.5	46.9	66.5	92.8	96.6	70.5	30.3	17.8	44.1	42.6	6.0	8.2
2003		74.4	46.8	76.6	90.5	92.4	69.4	31.5	16.0	44.6	43.3	6.2	8.6
2004[p]		74.7	45.4	67.5	86.3	87.8	69.0	30.9	14.5	44.9	43.6	6.0	8.2
2002	March	10.3	27.8	49.0	64.0	55.4	39.8	18.7	18.0	44.4	42.9	0.9	5.1
	June	19.7	52.7	60.8	98.8	106.6	79.1	34.4	17.3	44.2	42.7	1.7	9.2
	Sept	25.9	68.2	94.8	130.8	149.4	107.1	41.3	18.6	43.5	42.0	2.0	11.0
	Dec	14.6	38.5	61.2	76.8	74.0	55.6	26.4	16.9	44.7	43.3	1.3	7.2
2003	March	10.7	27.2	59.5	63.9	52.2	37.1	19.6	16.4	45.3	43.9	1.0	5.7
	June	21.0	53.0	74.9	94.3	105.2	77.4	36.5	15.8	44.8	43.5	1.8	10.0
	Sept	27.8	69.3	108.0	132.3	142.6	108.7	43.6	16.5	44.0	42.7	2.0	11.1
	Dec	14.9	37.2	63.5	70.9	68.7	53.6	26.3	15.2	45.1	43.8	1.4	7.5
2004	March[p]	10.5	25.8	67.9	58.3	50.5	37.3	17.9	15.4	45.2	43.7	1.0	5.7
	June[p]	20.7	50.7	56.9	85.7	94.1	76.6	35.3	13.7	45.2	43.8	1.7	9.6
	Sept[p]	28.1	67.9	83.4	124.9	137.7	107.3	44.0	14.9	44.4	43.1	2.0	10.8
	Dec[p]	15.4	37.2	61.7	75.8	68.6	54.7	26.2	14.2	45.4	44.0	1.2	6.7
Females													
1961		18.0	97.1	542.2	409.6	250.2	111.5	35.6	46.8	37.2	35.9	16.5	6.5
1966		25.1	114.7	567.8	411.2	254.8	135.9	37.8	52.4	36.2	34.3	16.8	6.3
1971		39.6	134.0	464.4	359.0	232.7	139.8	49.3	57.0	35.7	33.0	17.7	6.3
1976		65.1	122.2	458.9	272.3	188.0	124.0	40.9	59.8	34.9	32.4	17.0	5.9
1981		75.1	90.7	257.5	202.1	142.9	95.5	29.0	57.9	35.1	33.4	13.5	4.6
1986		80.0	68.7	190.9	155.9	111.6	75.6	24.4	51.2	36.0	34.7	11.2	3.8
1991		73.4	50.3	111.9	118.1	89.7	55.3	20.9	47.4	37.1	35.7	8.6	2.9
1994		76.9	47.3	167.3	121.0	91.4	54.4	20.6	44.4	37.9	36.3	7.9	2.7
1995		76.9	45.7	166.5	118.8	91.9	54.8	19.8	42.8	38.1	36.6	7.5	2.6
1996		78.9	45.6	183.5	120.6	93.6	56.0	20.4	40.8	38.6	37.1	7.3	2.6
1997		77.1	43.3	188.5	119.4	90.8	54.6	19.6	39.0	38.9	37.4	7.0	2.5
1998		73.3	40.1	175.0	114.5	87.1	52.2	18.4	37.1	39.3	37.9	6.6	2.4
1999		72.0	38.4	155.0	107.0	84.8	52.3	17.8	34.7	39.7	38.3	6.2	2.3
2000		74.1	38.5	137.8	107.5	85.6	54.2	18.4	32.0	40.1	38.9	6.2	2.3
2001		66.1	33.5	104.6	96.9	79.3	48.5	15.9	30.7	40.4	39.2	5.6	2.0
2002		69.2	34.3	107.5	101.2	81.7	51.2	16.9	28.2	40.9	39.7	5.7	2.1
2003		73.1	34.9	117.0	101.0	82.4	52.0	18.3	26.1	41.5	40.3	5.9	2.2
2004[p]		72.5	33.6	109.3	94.0	81.1	51.9	17.9	24.0	41.9	40.8	5.8	2.2
2002	March	10.4	20.9	77.7	72.2	49.6	30.1	10.3	29.8	40.8	39.6	0.9	1.3
	June	19.4	38.6	111.0	108.7	90.7	57.5	19.4	27.6	41.1	39.8	1.6	2.4
	Sept	24.9	49.0	139.6	141.4	120.5	75.0	22.9	28.6	40.6	39.5	1.9	2.8
	Dec	14.5	28.6	101.1	81.7	65.2	41.7	14.9	27.4	41.3	40.0	1.3	1.8
2003	March	10.9	21.1	95.5	69.6	50.3	29.7	11.3	27.7	41.6	40.2	0.9	1.4
	June	20.5	39.2	115.2	104.9	91.9	58.2	21.2	25.3	41.7	40.6	1.7	2.6
	Sept	26.6	50.3	138.5	147.1	120.6	78.1	24.9	26.2	41.2	40.1	2.0	2.9
	Dec	15.1	28.6	118.4	82.0	66.1	41.6	15.5	26.0	41.7	40.5	1.3	1.9
2004	March[p]	10.9	20.3	101.4	67.9	52.3	30.1	10.4	27.0	41.4	40.2	0.9	1.4
	June[p]	20.2	37.6	102.4	99.5	89.8	57.4	20.8	23.3	42.2	41.2	1.7	2.5
	Sept[p]	26.5	49.0	131.1	128.0	116.9	78.2	25.5	23.2	41.8	40.8	2.0	3.0
	Dec[p]	14.9	27.5	102.4	80.2	65.1	41.8	14.8	24.1	42.0	40.8	1.2	1.9

Notes: Marriage rates for 1986 have been calculated using the interim revised marital status estimates (based on the original mid-2001 population estimates) and are subject to further revision.

1 Figures for all marriages can be found in Table 2.1.
2 Per 1,000 divorced persons aged 16 and over.
3 The mean/median ages shown in this table are unstandardised and therefore take no account of changes in the structure of the population, by age or marital status.
4 Per 1,000 widowed persons aged 16 and over.
p provisional.

See 'Notes to tables'.

Table 9.3 **Divorces: age and sex**

England and Wales — Numbers (thousands), rates, percentages, mean and median age

Year and quarter	Petitions filed	Decrees made absolute			Divorce decrees per 1,000 married population						Per cent aged under 35	Mean age at divorce[1]	Median age at[1] divorce
		All divorces	1st marriage	2nd or later marriage	16 and over	16–24	25–29	30–34	35–44	45 and over			
		Numbers											
Males													
1961	13.7	25.4	23.5	1.9	2.1	1.4	3.9	4.1	3.1	1.1	38.3	..	..
1966	18.3	39.1	36.4	2.7	3.2	2.6	6.8	6.8	4.5	1.5	44.2	38.6	36.4
1971	44.2	74.4	69.3	5.2	5.9	5.0	12.5	11.8	7.9	3.1	44.8	39.4	36.6
1976	43.3	126.7	115.7	11.0	10.1	13.6	21.4	18.9	14.1	4.5	48.6	38.0	35.4
1981	46.7	145.7	127.6	18.1	11.9	17.7	27.6	22.8	17.0	4.8	48.6	37.7	35.4
1986	49.7	153.9	128.0	25.9	13.0	31.4	31.4	25.2	18.0	5.2	45.6	37.8	36.2
1991	..	158.7	129.8	29.0	13.6	26.1	32.4	28.6	20.2	5.6	42.7	38.6	37.0
1996	..	157.1	125.8	31.3	13.9	28.1	32.6	30.2	22.2	6.4	37.5	39.8	38.1
1997	..	146.7	117.3	29.4	13.1	26.0	30.4	28.7	21.1	6.1	35.9	40.2	38.4
1998	..	145.2	116.0	29.2	13.0	25.8	30.7	28.4	21.5	6.1	34.3	40.4	38.7
1999	..	144.6	115.1	29.4	13.0	24.1	29.7	28.4	21.9	6.3	32.1	40.9	39.2
2000	..	141.1	112.1	29.1	12.7	22.3	27.9	27.4	21.9	6.3	29.9	41.3	39.7
2001	..	143.8	114.3	29.5	13.0	20.3	27.9	28.3	22.8	6.5	28.4	41.5	40.0
2002	..	147.7	116.9	30.8	13.4	23.1	29.4	29.1	23.7	6.9	26.7	41.9	40.4
2003	..	153.5	121.4	32.0	14.0	24.6	30.6	29.8	25.2	7.4	24.7	42.3	40.9
2004[p]	..	153.4	121.1	32.3	14.1	25.5	30.4	29.9	25.3	7.7	23.3	42.7	41.4
2002 Sept	..	38.0	30.0	8.0	13.7	23.4	29.5	29.7	24.1	7.1	26.6	41.9	40.5
Dec	..	36.6	29.0	7.6	13.2	23.4	29.3	27.9	23.1	6.9	26.4	42.0	40.5
2003 March	..	39.4	31.2	8.2	14.6	26.1	33.5	31.4	26.0	7.7	25.3	42.1	40.7
June	..	38.6	30.4	8.1	14.1	23.7	30.3	30.7	25.5	7.4	25.0	42.2	40.9
Sept	..	37.9	30.0	7.9	13.8	24.7	29.5	28.8	24.8	7.3	24.4	42.3	41.0
Dec	..	37.6	29.7	7.8	13.6	23.8	29.0	28.5	24.4	7.3	24.3	42.4	41.1
2004 March[p]	..	39.5	31.2	8.3	14.6	25.4	31.7	31.9	26.5	7.8	23.8	42.5	41.2
June[p]	..	38.1	30.1	7.9	14.1	25.7	29.8	29.3	25.3	7.8	22.9	42.7	41.5
Sept[p]	..	39.0	30.9	8.1	14.3	25.8	30.7	30.2	25.6	7.8	23.2	42.7	41.5
Dec[p]	..	36.9	29.0	7.9	13.5	25.2	29.5	28.4	23.9	7.5	23.3	42.7	41.5
2005 March[p]	..	36.2	28.5	7.7	13.5	24.9	28.0	27.0	24.2	7.6	22.2	43.0	41.8
June[p]	..	36.5	28.7	7.8	13.5	23.7	27.5	26.0	23.8	7.8	21.5	43.2	42.1
Sept[p]	..	35.6	28.0	7.6	13.0	22.7	26.9	25.7	23.2	7.4	21.9	43.0	41.9
Females													
1961	18.2	25.4	23.4	2.0	2.1	2.4	4.5	3.8	2.7	0.9	49.3	..	..
1966	28.3	39.1	36.2	2.8	3.2	4.1	7.6	6.1	3.9	1.2	54.7	35.8	33.6
1971	66.7	74.4	69.3	5.1	5.9	7.5	13.0	10.5	6.7	2.8	54.4	36.8	33.6
1976	101.5	126.7	115.9	10.8	10.1	14.5	20.4	18.3	12.6	4.0	56.6	36.0	33.1
1981	123.5	145.7	127.7	18.0	11.9	22.3	26.7	20.2	14.9	3.9	58.0	35.2	33.2
1986	130.7	153.9	128.8	25.1	12.8	30.7	28.6	22.0	15.8	4.1	55.0	35.3	33.6
1991	..	158.7	130.9	27.8	13.4	28.7	30.7	25.0	17.3	4.5	52.7	36.0	34.3
1996	..	157.1	126.9	30.2	13.7	30.7	33.2	27.6	19.3	5.1	47.7	37.3	35.6
1997	..	146.7	118.3	28.4	12.9	28.0	31.3	26.3	18.5	4.9	45.9	37.7	36.0
1998	..	145.2	116.8	28.5	12.9	28.5	31.4	26.6	18.9	4.9	44.3	37.9	36.3
1999	..	144.6	115.4	29.1	12.9	25.6	30.6	26.9	19.5	5.1	41.7	38.4	36.9
2000	..	141.1	112.6	28.5	12.6	24.5	29.0	26.6	19.4	5.2	39.6	38.8	37.3
2001	..	143.8	114.6	29.2	12.9	23.9	29.2	27.6	20.5	5.4	37.8	39.1	37.7
2002	..	147.7	117.5	30.2	13.3	26.8	30.4	28.3	21.6	5.7	35.9	39.4	38.2
2003	..	153.5	121.9	31.6	14.0	28.2	31.6	29.1	23.2	6.1	33.7	39.8	38.7
2004[p]	..	153.4	121.8	31.6	14.0	27.9	32.0	28.8	23.6	6.4	31.9	40.2	39.2
2002 Sept	..	38.0	30.2	7.8	13.6	27.1	30.9	28.5	22.1	5.9	35.6	39.5	38.2
Dec	..	36.6	29.2	7.4	13.1	26.4	30.0	27.9	21.1	5.6	36.0	39.5	38.2
2003 March	..	39.4	31.3	8.1	14.5	30.1	33.7	30.9	23.9	6.3	34.4	39.7	38.5
June	..	38.6	30.7	7.9	14.1	28.0	31.5	29.6	23.4	6.1	33.7	39.8	38.7
Sept	..	37.9	30.0	8.0	13.7	28.0	30.2	28.3	23.1	6.0	33.3	39.8	38.8
Dec	..	37.6	29.9	7.6	13.5	26.8	31.0	27.6	22.4	6.1	33.3	39.9	38.9
2004 March[p]	..	39.5	31.4	8.1	14.5	28.8	33.6	30.7	24.5	6.5	32.6	40.0	39.0
June[p]	..	38.1	30.2	7.8	14.0	28.1	31.0	28.6	23.6	6.5	31.5	40.3	39.3
Sept[p]	..	39.0	30.9	8.1	14.2	27.8	32.2	28.9	23.8	6.6	31.7	40.3	39.3
Dec[p]	..	36.9	29.0	7.9	13.4	26.8	31.0	27.0	22.4	6.3	31.7	40.3	39.3
2005 March[p]	..	36.2	28.6	7.6	13.4	25.9	29.0	26.0	22.8	6.4	30.2	40.6	39.6
June[p]	..	36.5	28.8	7.7	13.4	26.2	27.7	25.9	22.7	6.5	29.8	40.7	39.9
Sept[p]	..	35.6	28.1	7.5	12.9	25.7	28.2	24.8	21.9	6.2	30.3	40.6	39.7

Notes: Divorce rates for 1986 have been calculated using the interim revised marital status estimates (based on the original mid-2001 estimates) and are subject to further revision.
1 The mean/median ages shown in this table are unstandardised and therefore take no account of changes in the structure of the population by age or marital status.
p provisional. See 'Notes to tables'.

Divorce petitions entered by year and quarter 1995–2005

England and Wales — Numbers (thousands)

Year	March Qtr	June Qtr	Sept Qtr	Dec Qtr	Year	March Qtr	June Qtr	Sept Qtr	Dec Qtr
1995	46.8	41.9	45.7	40.5	2001	45.4	42.6	42.9	42.0
1996	45.6	44.5	45.3	43.4	2002	45.4	44.3	45.4	42.6
1997	35.6	43.7	44.0	40.9	2003	46.3	42.2	43.6	41.5
1998	43.0	40.3	42.1	41.0	2004	45.5	41.1	42.1	39.1
1999	41.4	39.5	41.3	40.5	2005	37.9	39.5	38.5	36.1
2000	39.3	37.6	39.5	41.8					

1 Data supplied by Her Majesty's Court Service (14 February 2006) with the introduction of Management Information System Data (from 2001).
Note: The Divorce Reform Act 1969 became operative on 1 January 1971; the Matrimonial and Family Proceedings Act came into effect on 12 October 1984.
Figures include petitions for nullity

Notes to tables

Time Series
For most tables, years start at 1971 and then continue at five-year intervals until 1991. Individual years are shown thereafter.

United Kingdom
The United Kingdom comprises England, Wales, Scotland and Northern Ireland. The Channel Islands and the Isle of Man are not part of the United Kingdom.

Population
The estimated and projected resident population of an area includes all people who usually live there, whatever their nationality. Members of HM and US Armed Forces in the United Kingdom are included on a residential basis wherever possible. HM Forces stationed outside the United Kingdom are not included. Students are taken to be resident at their term-time addresses.

Live births
For England and Wales, figures relate to numbers occurring in a period; for Scotland and Northern Ireland, figures relate to those registered in a period.

Perinatal mortality
In October 1992 the legal definition of a stillbirth was changed, from baby born dead after 28 completed weeks of gestation or more, to one born dead after 24 completed weeks of gestation or more.

Expectation of life
The life tables on which these expectations are based use current death rates to describe mortality levels for each year. Each individual year shown is based on a three-year period, so that for instance 1986 represents 1985–87. More details can be found in *Population Trends* 60, page 23.

Deaths
Figures for England and Wales represent the numbers of deaths registered in each year up to 1992, and the number of deaths occurring in each year from 1993, though provisional figures are registrations. Figures for both Scotland and Northern Ireland represent the number of deaths registered in each year.

Age-standardised mortality
Directly age-standardised rates make allowances for changes in the age structure of the population. The age-standardised rate for a particular condition is that which would have occurred if the observed age-specific rates for the condition had applied in a given standard population. Table 2.2 uses the European Standard Population. This is a hypothetical population standard which is the same for both males and females allowing standardised rates to be compared for each sex, and between males and females.

International Migration
The UN recommends the following definition of an international long-term migrant.

An *international long-term migrant* is defined as a person who moves to a country other than that of his or her usual residence for a period of at least a year (12 months), so that the country of destination effectively becomes his or her new country of usual residence.

Figures in Tables 7.1–7.3 are compiled from several main sources of migration data:

- The richest source of information on international migrants comes from the International Passenger Survey (IPS), which is a sample survey of passengers arriving at, and departing from, the main United Kingdom air and sea ports and Channel Tunnel. This survey provides migration estimates based on respondents' *intended* length of stay in the UK or abroad and excludes most persons seeking asylum and some dependents of such asylum seekers.

- Two adjustments are made to account for people who do not realise their intended length of stay on arrival. First, visitor data from the IPS are used to estimate 'visitor switchers': those people who initially come to or leave the UK for a short period but subsequently stay for a year or longer. (For years before 2001, estimates of non-European Economic Area (non-EEA) national visitor switcher inflows are made from the Home Office database of after-entry applications to remain in the UK). Second, people who intend to be migrants, but who in reality stay in the UK or abroad for less than a year ('migrant switchers'), are estimated from IPS migrant data.

- Home Office data on asylum seekers and their dependents.

- Estimates of migration between the UK and the Irish Republic estimated using information from the Irish Quarterly National Household Survey and the National Health Service Central Register, agreed between the Irish Central Statistics Office and the ONS.

For years prior to 1991, the figures in Tables 7.1–7.3 are based only on data from the IPS. After taking into account of those groups of migrants known not to be covered by the IPS, it is estimated that the adjustment needed to net migration ranges from about ten thousand in 1981 to just over twenty thousand in 1986. From 1991, the figures in Tables 7.1–7.3 are based on data from all sources and represent Total International Migration.

Old Commonwealth is defined as Australia, Canada, New Zealand and South Africa;

New Commonwealth is defined as all other Commonwealth countries.

Middle East is defined as Bahrain, Iran, Iraq, Israel Jordan, Kuwait, Lebanon, Oman, Qatar, Saudi Arabia, Syria, the United Arab Emirates and Yemen.

Internal Migration
Figures in Table 8.1 are based on the movement of NHS doctors' patients between former Health Authorities (HAs) in England and Wales, and Area Health Boards in Scotland and Northern Ireland. Yearly and quarterly figures have been adjusted to take account of differences in recorded cross-border flows between England and Wales, Scotland and Northern Ireland.

Prior to reorganisation of health authority databases from Family Health Service Authorities (FHSAs) to HAs some database boundaries were realigned. This included in a few cases transferring patients between databases to fit the new boundaries. For the most part, this movement was done outside the NHSCR system and therefore had no effect on migration data. However a small number were transferred within the system. As migration estimates derived from NHSCR are the product of an administrative system (when patients re-register with GPs) this had the effect of generating small numbers of spurious migrants where no actual change of address had taken place. We have been advised of adjustments required to data by the Department of Health and these have been made to migration data.

The NHS Central Register (NHSCR) at Southport was computerised in early 1991, prior to which a three month time lag was assumed between a person moving and their re-registration with an NHS doctor being processed onto the NHSCR. Since computerisation, estimates of internal migration are based on the date of acceptance of the new patient by the HA (not previously available), and a one month time lag assumed.

It has been established that NHSCR data under-report the migration of males aged between 16 and 36. Currently, however, there are no suitable sources of data available to enable adjustments or revisions to be made to the estimates. Further research is planned on this topic and new data sources may become available in the future. However, for the present time, historical estimates will not be revised and future estimates will not be adjusted.

Marriages and divorces
Marriages are tabulated according to date of solemnisation. Divorces are tabulated according to date of decree absolute. In Scotland a small number of late divorces from previous years are added to the current year. The term 'divorces' includes decrees of nullity. The fact that a marriage or divorce has taken place in England, Wales, Scotland or Northern Ireland does not mean either of the parties is resident there.

EU Enlargement
The coverage of European countries in Table 1.1 has been updated to reflect the enlargement of the EU to 25 member countries (EU25) on 1 May 2004. The new member countries are: Cyprus, Czech Republic, Estonia, Hungary, Latvia, Lithuania, Malta, Poland, Slovakia and Slovenia. The main data source for these countries is the *United Nations Monthly Bulletin of Statistics*.

Sources
Figures for Scotland and Northern Ireland have been provided by the General Register Office for Scotland and the Northern Ireland Statistics and Research Agency respectively. The International Passenger Survey (Tables 7.1–7.3) is conducted by the Surveys and Administrative Sources Directorate of ONS.

Rounding
All figures are rounded independently; constituent parts may not add to totals. Generally numbers and rates per 1,000 population are rounded to one decimal place (e.g. 123.4); where appropriate, for small figures (below 10.0), two decimal places are given (e.g. 7.62). Figures which are provisional or estimated are given in less detail (e.g. 123 or 7.6 respectively) if their reliability does not justify giving the standard amount of detail. Where figures need to be treated with particular caution, an explanation is given as a footnote.

Latest figures
Figures for the latest quarters and years may be provisional and will be updated in future issues when later information becomes available. Where figures are not yet available, cells are left blank.

Shaded background
A shaded background indicates figures that are or may be subject to change: the grey shading signifies that the underlying estimates relate to those originally published; the coloured shading indicates estimates that have already been revised, from the original, but will or may be subject to further revision.

Report:

Live births in England and Wales, 2005: area of residence

This report provides summary statistics of live births in England and Wales during 2005 and compares them with figures for previous years. It also presents numbers of births and provisional fertility rates by area of residence of the mother. Further details of births in 2005 will be published in the volume *Births Statistics 2005* (Series FM1 no.34) on the National Statistics website (www.statistics.gov.uk).

KEY OBSERVATIONS

- The provisional general fertility rate (GFR) for 2005 was 58.5 live births per thousand women aged 15–44 – a slight increase on 2004 (58.2).

- There were 645,835 live births in England and Wales in 2005 compared with 639,721 in 2004 – an increase of 1 per cent.

- If the provisional 2005 patterns of fertility by age were to remain unchanged, as represented by the total fertility rate (TFR), then each woman would have an average of 1.80 children. This is the fourth consecutive annual increase from the low point in 2001 when the TFR was 1.63. The TFR is now at its highest level since 1992.

- In 2005, for all live births, the unstandardised average (mean) age of the mother was 29.5 years, an increase from 29.4 in 2004. This was one year older than the mean age of all mothers in 1995, but this difference partly reflects changes in the age structure of the population over the last 10 years. The standardised mean age of women giving birth is a measure that allows such comparisons to be made over time as it is not affected by changes in the age structure of the population during the period. In 2005, the standardised mean age of mothers increased to 29.0 years up from 28.9 in 2004 and 28.2 in 1995. The

Table 1	Summary of key live birth statistics

England and Wales

Year	Number of live births	Total fertility rate (TFR)[1]	General fertility rate (GFR): all births per 1,000 women aged 15–44	Sex ratio: male births per 1,000 female births	Mean age of mother at childbirth (years)[2]	Percentage of births outside marriage	Percentage of births to non-UK born mothers
1995	648,138	1.72	60.5	1,051	28.5	33.9	12.6
1996	649,485	1.74	60.6	1,055	28.6	35.8	12.8
1997	643,095	1.73	60.0	1,051	28.8	37.0	13.1
1998	635,901	1.72	59.2	1,051	28.9	37.8	13.6
1999	621,872	1.70	57.8	1,055	29.0	38.9	14.3
2000	604,441	1.65	55.9	1,050	29.1	39.5	15.5
2001	594,634	1.63	54.7	1,050	29.2	40.0	16.5
2002	596,122	1.65	54.7	1,055	29.3	40.6	17.7
2003	621,469	1.73	56.8	1,051	29.4	41.4	18.6
2004	639,721	1.78	58.2	1,054	29.4	42.2	19.5
2005	645,835	1.80[3]	58.5[3]	1,049	29.5	42.8	20.8

1 The total fertility rate is the average number of children that would be born per woman if women experienced the age-specific fertility rates of the year in question throughout their childbearing lifespan.
2 The mean is unstandardised, showing the mean age of the mothers who had births in 2005, and is not adjusted for the age structure of the population.
3 Fertility rates are provisional as they are calculated using the 2004-based population projections for 2005.

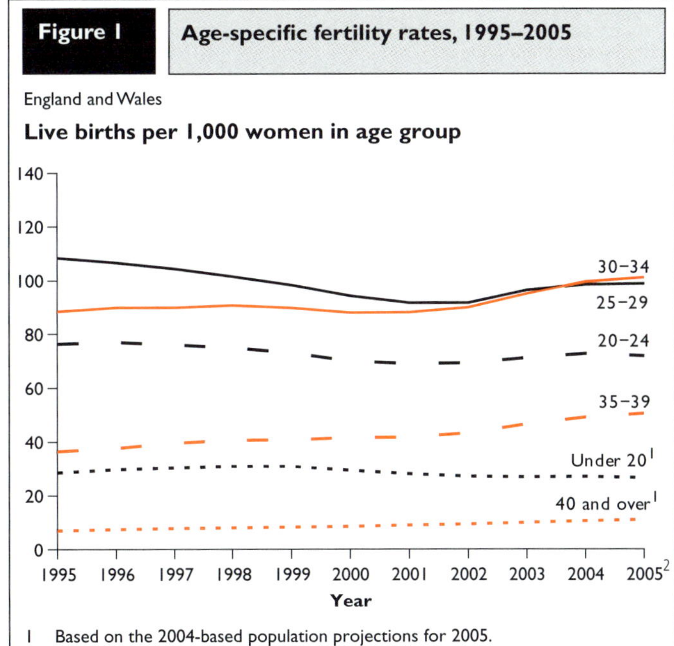

Figure 1	Age-specific fertility rates, 1995–2005

England and Wales

Live births per 1,000 women in age group

1 Based on the 2004-based population projections for 2005.
2 The rates for women aged under 20, and 40 and over, are based upon the population of women aged 15–19 and 40–44 respectively.

standardised average age of mothers at birth in 2005 was 2.6 years higher than when the lowest average age was seen in 1974 (26.4).

- In 2005, there were increases in age-specific fertility rates (ASFRs) for women in age groups 25 and over. The age-specific fertility rates for women aged under 25 have decreased. Provisional calculations show that women aged 30 to 34 had the highest fertility rate, at 100.9 births per thousand women, an increase of 1.5 per cent from 99.4 births per thousand women in 2004.

- The fertility rate for women aged 30 to 34 first overtook that of women in the age group 25 to 29 in 2004 and remained higher in 2005. This is the first time since 1998 that the fertility rates for any age group exceeded 100 births per thousand women, and not since 1965 have fertility rates for women aged 30 to 34 exceeded 100 births per thousand women. In 2005, fertility increased fastest for women aged 40 and over, rising to 10.8 births per thousand women – an increase of 4 per cent compared with 2004 (10.4).

- The long-term rise in the proportion of births outside marriage continued: 42.8 per cent of live births were outside marriage in 2005, compared with 42.2 per cent in 2004 and 33.9 per cent in 1995.

- The proportion of live births to mothers born outside the United Kingdom continued to rise. Over 1 in 5 births in 2005 were to mothers born outside the United Kingdom compared to 1 in 8 in 1995. The number of live births to mothers born outside the United Kingdom increased by nearly 8 per cent from 124,563 births in 2004 to 134,189 in 2005. There were 81,677 births to mothers born outside the United Kingdom in 1995 – an increase of 64 per cent from 1995 to 2005.

- The increase in the number of births in England and Wales to mothers born outside the United Kingdom is due partly to the increase in mothers born in other European countries (that is, from both the European Union and the rest of Europe). In 1995, 13,473 live births in England and Wales were to mothers born in other European countries. This was 2.1 per cent of all live births. In 2005, the proportion of births to mothers born in other European countries had more than doubled to 4.3 per cent (27,924) of live births. See explanatory note 5.

- The number of births to mothers born in countries belonging to the European Union also increased. In 1995, there were 11,113 live births in England and Wales to mothers born in one of the other 23 member countries of the European Union (excluding the United Kingdom and the Republic of Ireland), as constituted in 2005. This was 1.7 per cent of all live births that year. In 2005, this proportion almost doubled to 3.2 per cent, with 20,420 births.

Variations in fertility by area are shown in Tables 2 and 3 where numbers of births, and provisional GFRs and TFRs are presented for administrative and health areas.

- The highest level of fertility among the Government Office Regions of England in 2005, as represented by the GFR, was in London with 63.8 live births per thousand women aged 15–44, followed by the West Midlands (61.1). The lowest GFR was in the North East (54.4).

- The highest fertility level among the Government Office Regions of England in 2005, as represented by the TFR was in the West Midlands where there was an average of 1.91 children per woman. The lowest was in the North East with 1.74.

- Two London boroughs recorded the highest GFRs in England and Wales. The borough of Newham was highest with 86.8 live births per thousand women aged 15–44, followed by Barking and Dagenham (79.2). The lowest GFR was in Cambridge county district (37.4).

- Newham had the highest TFR of 2.47 children per woman, followed by Blackburn with Darwen unitary authority with a TFR of 2.43. The lowest TFR was in Kensington and Chelsea (1.19) followed by Cambridge county district (1.21). See explanatory note 3.

- Among the Strategic Health Authorities in England, North East London had the highest GFR with 73.1 live births per thousand women aged 15–44 and the highest TFR with 2.08 children per woman. Northumberland, Tyne and Wear had the lowest GFR at 52.4 and South West London had the lowest TFR at 1.62.

- In Wales, the national GFR was 56.1 live births per thousand 15–44 year-old women and the TFR was 1.79 children per woman.

- In Wales, the Local Health Board and unitary authority with the highest GFR (62.4) and TFR (2.07) was Torfaen while the lowest for both fertility rates was Ceredigion where the GFR was 38.5 live births per thousand women and the TFR was 1.38.

EXPLANATORY NOTES

1. In this report, all fertility rates and standardised mean ages of mothers for 2005 are provisional. At subnational level they have been calculated using mid-2004 population estimates and at the national level the 2004-based population projections for 2005 have been used. The population projections are available on the Government Actuary's Department website (www.gad.gov.uk). The population figures used to calculate national level fertility rates for 2004 and earlier years are ONS mid-year population estimates. The population estimates used were the most up-to-date at the time of publication of this report. Further information on population estimates can be found on the National Statistics website (www.statistics.gov.uk/popest).

2. Numbers of births, GFRs and TFRs are given by mother's usual area of residence, based on 2005 Local and Strategic Health Authority area boundaries (Local Health Boards in Wales).

3. The TFR has been calculated using the number of births and 2004 mid-year population estimates for women by single year of age. This generally produces a better match of births to those at risk of having births. However, local authority level population estimates are only considered reliable in five-year age bands. Thus, especially in small local authorities, it should be noted that rates computed using single year of age data may produce spurious results. In particular the rate for Rutland is affected by the very low estimated population aged 19, 20 and 21. Computing the rate for Rutland using five-year age band data gives a TFR of 2.15 compared to 3.44 which is computed when using single year of age data.

4. The unstandardised mean age shows the mean age of mothers who had births in 2005 and is not adjusted for the age structure of the population. The provisional standardised mean age is a measure which allows fertility trends to be separated out from the effects of changes in the population's age structure over time.

5. For comparability, the 1995 births data for mothers born outside the United Kingdom were reclassified according to the 2005 country classification list and the definition of the European Union, as constituted in 2005, was used for both years' data. The percentage of births to mothers born in Europe excludes births where the mother was born in the United Kingdom or the Republic of Ireland.

6. Further information on live births in 2005 can be found in Reference Tables 2.1 and 3.1 to 3.3 in this publication and at the births topic-based summary on the National Statistics website: www.statistics.gov.uk/cci/nugget.asp?id=369.

Table 2 — Live births by local authority of usual residence of mother, numbers, general fertility rates and total fertility rates, 2005

England and Wales, Government Office Regions (within England), unitary authorities/counties/districts & London boroughs

Area of usual residence	Live births	GFR[1]	TFR[2,3]
ENGLAND AND WALES	**645,835**	**58.5**	**1.80**
ENGLAND	**613,028**	**58.6**	**1.80**
NORTH EAST	**28,249**	**54.4**	**1.74**
Darlington UA	1,218	62.0	2.02
Hartlepool UA	1,116	61.1	2.03
Middlesbrough UA	1,916	65.2	2.07
Redcar and Cleveland UA	1,577	58.1	1.95
Stockton-on-Tees UA	2,241	57.4	1.86
Durham	**5,189**	**52.2**	**1.69**
Chester-le-Street	537	51.4	1.73
Derwentside	922	54.9	1.80
Durham	788	37.8	1.33
Easington	1,069	58.1	1.91
Sedgefield	958	55.7	1.82
Teesdale	193	47.3	1.68
Wear Valley	722	61.5	2.13
Northumberland	**3,024**	**53.3**	**1.80**
Alnwick	296	54.0	1.91
Berwick-upon-Tweed	194	44.6	1.60
Blyth Valley	975	60.0	1.91
Castle Morpeth	368	44.7	1.54
Tynedale	516	49.6	1.73
Wansbeck	675	56.2	1.89
Tyne and Wear (Met County)	**11,968**	**52.2**	**1.63**
Gateshead	2,118	54.6	1.74
Newcastle upon Tyne	2,979	47.4	1.48
North Tyneside	2,274	59.8	1.92
South Tyneside	1,529	50.3	1.67
Sunderland	3,068	51.7	1.62
NORTH WEST	**81,722**	**58.1**	**1.82**
Blackburn with Darwen UA	2,284	76.9	2.43
Blackpool UA	1,649	60.9	1.99
Halton UA	1,650	65.2	2.07
Warrington UA	2,207	55.3	1.76
Cheshire	**7,317**	**55.3**	**1.76**
Chester	1,278	51.9	1.57
Congleton	866	49.8	1.61
Crewe and Nantwich	1,317	59.5	1.94
Ellesmere Port & Neston	912	57.7	1.89
Macclesfield	1,490	52.9	1.66
Vale Royal	1,454	60.1	2.00
Cumbria	**4,784**	**52.4**	**1.72**
Allerdale	895	51.5	1.71
Barrow-in-Furness	750	55.5	1.85
Carlisle	1,107	54.2	1.70
Copeland	724	53.7	1.76
Eden	456	49.5	1.61
South Lakeland	852	49.1	1.67
Greater Manchester (Met County)	**33,544**	**61.5**	**1.87**
Bolton	3,576	65.0	2.05
Bury	2,283	61.1	1.94
Manchester	6,707	60.0	1.71
Oldham	3,200	71.6	2.29
Rochdale	2,853	66.7	2.13
Salford	2,915	63.6	1.90
Stockport	3,151	55.8	1.76
Tameside	2,533	56.3	1.82
Trafford	2,631	59.7	1.83
Wigan	3,695	59.0	1.87

– continued	Live births	GFR[1]	TFR[2,3]
Lancashire	**13,084**	**57.0**	**1.83**
Burnley	1,163	64.2	2.08
Chorley	1,177	58.3	1.86
Fylde	655	50.7	1.71
Hyndburn	1,123	68.9	2.20
Lancaster	1,401	47.8	1.55
Pendle	1,244	70.0	2.24
Preston	1,801	60.6	1.85
Ribble Valley	475	45.6	1.52
Rossendale	803	60.6	1.99
South Ribble	1,114	53.0	1.73
West Lancashire	1,193	55.9	1.85
Wyre	935	48.5	1.66
Merseyside (Met County)	**15,203**	**53.0**	**1.68**
Knowsley	1,915	58.7	1.98
Liverpool	5,188	49.8	1.52
Sefton	2,584	48.1	1.61
St Helens	1,925	54.3	1.75
Wirral	3,591	59.1	1.96
YORKSHIRE AND THE HUMBER	**60,665**	**58.5**	**1.84**
East Riding of Yorkshire UA	2,887	49.7	1.67
Kingston upon Hull, City of UA	3,203	59.6	1.79
North East Lincolnshire UA	1,927	61.4	2.07
North Lincolnshire UA	1,796	60.7	2.04
York UA	1,929	46.7	1.44
North Yorkshire	**5,700**	**54.3**	**1.83**
Craven	456	50.2	1.73
Hambleton	803	54.1	1.90
Harrogate	1,626	55.5	1.76
Richmondshire	527	55.7	1.79
Ryedale	425	49.4	1.65
Scarborough	1,022	55.4	1.92
Selby	841	55.3	1.94
South Yorkshire (Met County)	**15,081**	**57.2**	**1.79**
Barnsley	2,505	56.6	1.87
Doncaster	3,554	62.6	2.08
Rotherham	2,921	57.7	1.86
Sheffield	6,101	54.4	1.67
West Yorkshire (Met County)	**28,142**	**62.1**	**1.90**
Bradford	8,014	78.2	2.37
Calderdale	2,486	63.3	2.10
Kirklees	5,309	64.9	2.02
Leeds	8,709	53.0	1.61
Wakefield	3,624	55.2	1.79
EAST MIDLANDS	**49,080**	**56.7**	**1.79**
Derby UA	2,989	59.6	1.80
Leicester UA	4,597	67.6	1.97
Nottingham UA	3,753	55.2	1.65
Rutland UA[3]	344	55.9	3.44
Derbyshire	**7,902**	**55.2**	**1.80**
Amber Valley	1,197	52.9	1.71
Bolsover	890	61.9	2.06
Chesterfield	1,093	55.8	1.81
Derbyshire Dales	563	48.9	1.66
Erewash	1,227	54.7	1.77
High Peak	965	54.5	1.79
North East Derbyshire	881	50.4	1.68
South Derbyshire	1,086	62.0	1.96

Note: Rates are provisional - at subnational level, they are based on the most up-to-date mid-2004 population estimates released in December 2005. At national level, however, they are based on the 2004-based population projections for 2005.

1. The general fertility rate (GFR) is the number of live births per 1,000 women aged 15–44.

2. The total fertility rate (TFR) is the average number of children that would be born per woman if women experienced the age-specific fertility rates of the year in question throughout their childbearing lifespan.

3. The TFR has been calculated using the 2004 mid-year population estimates for women by single year of age. While generally calculating the rates using single age figures produces a better match of births to those at risk of having births, it should be noted that local authority level population estimates are only considered reliable down to five-year age bands. Thus, especially in small local authorities, the overall rates computed using single year of age data may produce spurious results. In particular the rate for Rutland is affected by the very low estimated population aged 19, 20 and 21. Computing the rate for Rutland using five-year age band data gives a TFR of 2.15.

Table 2 continued	Live births by local authority of usual residence of mother, numbers, general fertility rates and total fertility rates, 2005

England and Wales, Government Office Regions (within England), unitary authorities/counties/districts & London boroughs

Area of usual residence	Live births	GFR[1]	TFR[2]	- continued	Live births	GFR[1]	TFR[2]
Leicestershire	**6,657**	**54.1**	**1.74**	**West Midlands (Met County)**	**36,038**	**66.0**	**2.01**
Blaby	1,021	57.2	1.82	Birmingham	15,893	71.1	2.11
Charnwood	1,645	49.4	1.58	Coventry	3,871	59.3	1.78
Harborough	858	58.0	1.89	Dudley	3,546	59.7	1.90
Hinckley and Bosworth	1,028	52.2	1.67	Sandwell	4,171	69.5	2.16
Melton	508	55.5	1.88	Solihull	2,028	53.0	1.74
North West Leicestershire	1,024	60.0	1.92	Walsall	3,424	68.3	2.19
Oadby and Wigston	573	51.1	1.85	Wolverhampton	3,105	62.6	1.91
Lincolnshire	**6,671**	**53.9**	**1.79**	**Worcestershire**	**6,041**	**57.4**	**1.82**
Boston	683	66.6	2.22	Bromsgrove	876	52.4	1.71
East Lindsey	1,186	54.6	1.91	Malvern Hills	618	50.3	1.81
Lincoln	1,045	51.9	1.62	Redditch	1,053	63.4	1.94
North Kesteven	947	50.2	1.65	Worcester	1,322	64.1	1.90
South Holland	750	54.4	1.83	Wychavon	1,184	56.8	1.82
South Kesteven	1,253	52.2	1.75	Wyre Forest	988	54.2	1.72
West Lindsey	807	53.9	1.93	**EAST**	**64,687**	**59.4**	**1.85**
Northamptonshire	**8,279**	**62.6**	**2.01**	**Luton UA**	3,196	78.6	2.35
Corby	677	61.8	2.14	**Peterborough UA**	2,446	72.4	2.25
Daventry	857	59.6	1.97	**Southend-on-Sea UA**	1,948	62.8	1.96
East Northamptonshire	948	60.0	2.06	**Thurrock UA**	2,211	68.7	2.07
Kettering	1,156	67.1	2.11				
Northampton	2,828	65.2	1.98	**Bedfordshire**	**4,881**	**60.3**	**1.88**
South Northamptonshire	883	54.4	1.77	Bedford	1,922	61.1	1.84
Wellingborough	930	64.8	2.10	Mid Bedfordshire	1,515	58.1	1.78
				South Bedfordshire	1,444	61.8	2.02
Nottinghamshire	**7,888**	**52.5**	**1.67**	**Cambridgeshire**	**6,505**	**53.4**	**1.61**
Ashfield	1,282	55.0	1.77	Cambridge	1,189	37.4	1.21
Bassetlaw	1,118	53.4	1.81	East Cambridgeshire	964	63.5	2.02
Broxtowe	1,039	47.0	1.44	Fenland	917	58.4	2.00
Gedling	1,109	50.3	1.58	Huntingdonshire	1,855	56.9	1.80
Mansfield	1,143	57.1	1.86	South Cambridgeshire	1,580	59.8	1.80
Newark and Sherwood	1,111	53.5	1.77	**Essex**	**14,905**	**58.0**	**1.82**
Rushcliffe	1,086	51.3	1.56	Basildon	2,208	63.4	1.96
WEST MIDLANDS	**65,956**	**61.1**	**1.91**	Braintree	1,663	62.3	1.96
				Brentwood	712	53.6	1.66
Herefordshire, County of UA	1,654	53.1	1.76	Castle Point	827	53.6	1.81
Stoke-on-Trent UA	3,312	66.6	2.06	Chelmsford	1,850	56.8	1.72
Telford and Wrekin UA	2,057	60.3	1.94	Colchester	1,906	56.8	1.72
				Epping Forest	1,400	58.5	1.79
Shropshire	**2,767**	**54.5**	**1.81**	Harlow	1,109	65.5	2.01
Bridgnorth	425	47.7	1.58	Maldon	587	54.1	1.76
North Shropshire	611	59.0	1.97	Rochford	746	51.2	1.66
Oswestry	391	54.3	1.81	Tendring	1,146	52.9	1.85
Shrewsbury and Atcham	1,016	57.3	1.88	Uttlesford	751	58.4	1.90
South Shropshire	324	49.4	1.71				
Staffordshire	**8,503**	**54.3**	**1.78**	**Hertfordshire**	**12,975**	**60.0**	**1.82**
Cannock Chase	1,108	57.4	1.88	Broxbourne	1,048	58.9	1.84
East Staffordshire	1,271	61.2	2.06	Dacorum	1,623	58.6	1.82
Lichfield	982	57.4	1.88	East Hertfordshire	1,595	58.2	1.71
Newcastle-under-Lyme	1,236	49.4	1.60	Hertsmere	1,211	62.5	1.94
South Staffordshire	906	48.1	1.67	North Hertfordshire	1,418	58.5	1.78
Stafford	1,253	54.5	1.77	St Albans	1,769	65.2	1.85
Staffordshire Moorlands	861	51.2	1.70	Stevenage	1,040	60.0	1.90
Tamworth	886	56.8	1.80	Three Rivers	900	54.3	1.67
				Watford	1,148	64.3	1.84
Warwickshire	**5,584**	**53.3**	**1.65**	Welwyn Hatfield	1,223	58.8	1.84
North Warwickshire	622	51.3	1.67				
Nuneaton and Bedworth	1,453	60.2	1.93	**Norfolk**	**8,127**	**54.6**	**1.74**
Rugby	1,059	59.9	1.90	Breckland	1,258	56.2	1.83
Stratford-on-Avon	1,025	48.5	1.52	Broadland	1,088	50.7	1.64
Warwick	1,425	47.9	1.42	Great Yarmouth	995	59.8	1.98
				King's Lynn and West Norfolk	1,432	59.8	1.98
				North Norfolk	801	53.4	1.86

1   The general fertility rate (GFR) is the number of live births per 1,000 women aged 15–44.

2   The total fertility rate (TFR) is the average number of children that would be born per woman if woman experienced the age-specific fertility rates of the year in question throughout their childbearing lifespan.

Table 2 continued	Live births by local authority of usual residence of mother, numbers, general fertility rates and total fertility rates, 2005

England and Wales, Government Office Regions (within England), unitary authorities/counties/districts & London boroughs

Area of usual residence	Live births	GFR[1]	TFR[2]	- continued	Live births	GFR[1]	TFR[2]
Norwich	1,543	52.2	1.51	**Windsor and Maidenhead UA**	1,699	60.7	1.79
South Norfolk	1,010	50.8	1.69	**Wokingham UA**	1,694	53.0	1.61
Suffolk	**7,493**	**59.4**	**1.91**	**Buckinghamshire**	**5,511**	**58.1**	**1.80**
Babergh	863	58.2	1.97	Aylesbury Vale	1,956	56.6	1.76
Forest Heath	756	58.9	1.65	Chiltern	832	51.7	1.67
Ipswich	1,611	66.2	1.99	South Bucks	615	53.1	1.65
Mid Suffolk	904	57.9	1.96	Wycombe	2,108	64.8	1.97
St Edmundsbury	1,206	63.6	1.99				
				East Sussex	**4,783**	**56.9**	**1.92**
Suffolk Coastal	1,006	50.8	1.78	Eastbourne	968	57.0	1.79
Waveney	1,147	58.0	1.98	Hastings	1,014	62.4	2.03
				Lewes	870	55.8	1.88
LONDON	**116,019**	**63.8**	**1.79**	Rother	644	52.6	1.93
				Wealden	1,287	55.8	1.96
Inner London	**49,548**	**62.2**	**1.71**				
Camden	2,954	48.0	1.32	**Hampshire**	**13,537**	**55.5**	**1.77**
Hackney plus City of London[3]	4,439	77.0	2.20	Basingstoke and Deane	1,920	58.8	1.80
Hammersmith and Fulham	2,686	53.9	1.49	East Hampshire	1,142	55.5	1.87
Haringey	4,026	69.1	1.97	Eastleigh	1,298	55.5	1.75
				Fareham	961	47.4	1.60
Islington	2,731	53.4	1.49	Gosport	942	58.8	1.84
Kensington and Chelsea	2,188	43.9	1.19				
Lambeth	4,739	65.7	1.87	Hart	1,028	59.6	1.79
Lewisham	4,284	68.5	2.00	Havant	1,157	54.4	1.87
Newham	5,353	86.8	2.47	New Forest	1,519	53.1	1.80
				Rushmoor	1,238	61.3	1.81
Southwark	4,714	71.3	2.01	Test Valley	1,189	54.9	1.77
Tower Hamlets	3,968	69.5	1.82	Winchester	1,143	51.8	1.66
Wandsworth	4,554	54.8	1.51				
Westminster	2,912	44.6	1.23	**Kent**	**15,613**	**59.0**	**1.89**
				Ashford	1,322	62.6	2.05
Outer London	**66,471**	**64.9**	**1.89**	Canterbury	1,415	48.1	1.58
Barking and Dagenham	2,985	79.2	2.36	Dartford	1,242	67.1	2.08
Barnet	4,728	62.8	1.77	Dover	1,151	58.9	1.99
Bexley	2,686	58.6	1.85	Gravesham	1,203	63.0	2.03
Brent	4,503	69.2	1.96				
Bromley	3,663	59.1	1.76	Maidstone	1,658	59.2	1.82
				Sevenoaks	1,232	60.1	1.89
Croydon	4,704	60.0	1.80	Shepway	1,066	59.6	1.98
Ealing	4,838	66.9	1.89	Swale	1,497	60.4	1.97
Enfield	4,496	71.4	2.16	Thanet	1,416	61.5	2.00
Greenwich	3,963	71.6	2.03				
Harrow	2,872	62.2	1.83	Tonbridge and Malling	1,282	58.4	1.86
				Tunbridge Wells	1,129	54.9	1.77
Havering	2,474	55.9	1.77				
Hillingdon	3,489	61.2	1.81	**Oxfordshire**	**7,575**	**56.8**	**1.70**
Hounslow	3,674	73.3	2.09	Cherwell	1,807	64.6	2.01
Kingston upon Thames	2,000	55.7	1.59	Oxford	1,782	44.7	1.42
Merton	2,925	63.3	1.73	South Oxfordshire	1,544	62.1	1.87
				Vale of White Horse	1,359	60.9	1.91
Redbridge	3,577	66.0	1.94	West Oxfordshire	1,083	59.3	1.90
Richmond upon Thames	2,580	60.6	1.59				
Sutton	2,325	60.4	1.82	**Surrey**	**12,303**	**57.2**	**1.72**
Waltham Forest	3,989	74.1	2.13	Elmbridge	1,629	62.2	1.79
				Epsom and Ewell	782	57.3	1.73
SOUTH EAST	**93,921**	**57.5**	**1.77**	Guildford	1,436	50.7	1.52
				Mole Valley	789	55.4	1.73
Bracknell Forest UA	1,431	57.2	1.73	Reigate and Banstead	1,491	59.2	1.77
Brighton and Hove UA	3,045	50.0	1.42				
Isle of Wight UA	1,188	50.8	1.72	Runnymede	840	47.4	1.47
Medway UA	3,134	58.3	1.84	Spelthorne	1,079	60.6	1.88
Milton Keynes UA	3,200	67.4	2.06	Surrey Heath	920	56.5	1.76
				Tandridge	862	58.6	1.91
Portsmouth UA	2,333	53.6	1.59	Waverley	1,223	56.4	1.75
Reading UA	2,146	62.0	1.75	Woking	1,252	65.1	1.92
Slough UA	2,108	77.0	2.24				
Southampton UA	2,775	53.4	1.54				
West Berkshire UA	1,796	61.8	1.98				

1 The general fertility rate (GFR) is the number of live births per 1,000 women aged 15–44.

2 The total fertility rate (TFR) is the average number of children that would be born per woman if woman experienced the age-specific fertility rates of the year in question throughout their childbearing lifespan.

3 To protect confidentiality, all births for the City of London LB and Isles of Scilly CD have been included with those for Hackney LB and Penwith CD respectively.

Table 2 continued	Live births by local authority of usual residence of mother, numbers, general fertility rates and total fertility rates, 2005

England and Wales, Government Office Regions (within England), unitary authorities/counties/districts & London boroughs

Area of usual residence	Live births	GFR[1]	TFR[2]	- continued	Live births	GFR[1]	TFR[2]
West Sussex	**8,050**	**57.6**	**1.83**	**Gloucestershire**	**5,946**	**53.8**	**1.73**
Adur	584	55.5	1.84	Cheltenham	1,126	47.6	1.45
Arun	1,342	57.3	1.91	Cotswold	739	50.5	1.65
Chichester	988	52.7	1.70	Forest of Dean	773	53.4	1.84
Crawley	1,305	60.5	1.80	Gloucester	1,441	61.6	1.99
Horsham	1,293	56.1	1.76	Stroud	1,068	54.1	1.80
				Tewkesbury	799	54.4	1.77
Mid Sussex	1,423	58.9	1.83				
Worthing	1,115	61.3	1.94	**Somerset**	**5,149**	**56.1**	**1.88**
				Mendip	1,113	56.0	1.89
SOUTH WEST	**52,729**	**55.2**	**1.75**	Sedgemoor	1,075	55.7	1.91
				South Somerset	1,579	57.9	1.95
Bath and North East Somerset UA	1,702	48.2	1.53	Taunton Deane	1,130	56.2	1.79
Bournemouth UA	1,619	47.6	1.39	West Somerset	252	48.1	1.78
Bristol, City of UA	5,440	57.7	1.68				
North Somerset UA	2,032	59.3	1.95	**Wiltshire**	**4,835**	**57.3**	**1.86**
Plymouth UA	2,813	55.0	1.73	Kennet	839	59.3	2.02
				North Wiltshire	1,493	59.2	1.90
Poole UA	1,461	56.8	1.82	Salisbury	1,219	56.2	1.82
South Gloucestershire UA	2,955	58.8	1.84	West Wiltshire	1,284	54.9	1.78
Swindon UA	2,480	63.3	1.97				
Torbay UA	1,328	58.2	1.95	**WALES**	**32,593**	**56.1**	**1.79**
Cornwall and Isles of Scilly	**4,789**	**52.7**	**1.73**	Isle of Anglesey	685	56.8	1.87
Caradon	715	50.9	1.76	Gwynedd	1,264	57.3	1.79
Carrick	758	46.4	1.48	Conwy	1,044	55.2	1.85
Kerrier	976	56.8	1.89	Denbighshire	984	57.7	1.92
North Cornwall	761	53.8	1.80	Flintshire	1,642	55.3	1.77
Penwith plus Isles of Scilly[3]	563	50.8	1.68				
				Wrexham	1,582	61.1	1.90
Restormel	1,016	56.3	1.83	Powys	1,248	57.0	1.90
				Ceredigion	600	38.5	1.38
Devon	**6,693**	**52.6**	**1.74**	Pembrokeshire	1,201	58.5	1.96
East Devon	1,024	52.2	1.80	Carmarthenshire	1,744	53.9	1.77
Exeter	1,252	46.6	1.40				
Mid Devon	786	61.1	2.08	Swansea	2,449	54.6	1.73
North Devon	886	57.8	1.94	Neath Port Talbot	1,486	57.4	1.92
South Hams	648	48.4	1.77	Bridgend	1,523	59.5	1.97
				The Vale of Glamorgan	1,279	53.8	1.76
Teignbridge	1,127	53.3	1.84	Cardiff	3,955	51.8	1.56
Torridge	589	58.4	2.07				
West Devon	381	48.2	1.81	Rhondda, Cynon, Taff	2,864	60.7	1.92
				Merthyr Tydfil	643	58.2	1.97
Dorset	**3,487**	**54.4**	**1.90**	Caerphilly	2,055	59.9	1.91
Christchurch	372	55.2	1.96	Blaenau Gwent	734	54.1	1.84
East Dorset	675	52.7	1.94	Torfaen	1,094	62.4	2.07
North Dorset	643	59.1	2.06				
Purbeck	343	45.6	1.57	Monmouthshire	819	53.4	1.84
West Dorset	803	54.4	1.97	Newport	1,698	59.8	1.96
Weymouth and Portland	651	57.6	1.91				
				Normal residence outside England and Wales	**214**	:	:

1 The general fertility rate (GFR) is the number of live births per 1,000 women aged 15–44.

2 The total fertility rate (TFR) is the average number of children that would be born per woman if women experienced the age-specific fertility rates of the year in question throughout their childbearing lifespan.

3 To protect confidentiality, all births for the City of London LB and Isles of Scilly CD have been included with those for Hackney LB and Penwith CD respectively.

Table 3	Live births by health area of usual residence of mother, numbers, general fertility rates and total fertility rates, 2005

England and Wales, Government Office Regions (within England), and health authorities/boards[1]

Area of usual residence	Live births	GFR[2]	TFR[3]	- continued	Live births	GFR[2]	TFR[3]
ENGLAND AND WALES	**645,835**	**58.5**	**1.80**	**SOUTH EAST**	**93,921**	**57.5**	**1.77**
ENGLAND	**613,028**	**58.6**	**1.80**	Hampshire and Isle of Wight	19,833	54.7	1.70
				Kent and Medway	18,747	58.9	1.88
NORTH EAST	**28,249**	**54.4**	**1.74**	Surrey and Sussex	28,181	56.4	1.73
				Thames Valley	27,160	60.1	1.80
County Durham and Tees Valley	13,257	56.9	1.84				
Northumberland, Tyne & Wear	14,992	52.4	1.65	**SOUTH WEST**	**52,729**	**55.2**	**1.75**
NORTH WEST	**81,722**	**58.1**	**1.82**	Avon, Gloucestershire and Wiltshire	25,390	56.6	1.75
				Dorset and Somerset	11,716	54.3	1.77
Cheshire & Merseyside	26,377	54.5	1.74	South West Peninsula	15,623	53.5	1.75
Cumbria and Lancashire	21,801	57.7	1.86				
Greater Manchester	33,544	61.5	1.87	**WALES**	**32,593**	**56.1**	**1.79**
YORKSHIRE AND THE HUMBER	**60,665**	**58.5**	**1.84**	Anglesey	685	56.8	1.87
				Gwynedd	1,264	57.3	1.79
North and East Yorkshire and				Conwy	1,044	55.2	1.85
Northern Lincolnshire	17,442	54.7	1.77	Denbighshire	984	57.7	1.92
South Yorkshire	15,081	57.2	1.79	Flintshire	1,642	55.3	1.77
West Yorkshire	28,142	62.1	1.90				
				Wrexham	1,582	61.1	1.90
EAST MIDLANDS	**49,080**	**56.7**	**1.79**	Powys	1,248	57.0	1.90
				Ceredigion	600	38.5	1.38
Leicestershire, Northamptonshire				Pembrokeshire	1,201	58.5	1.96
and Rutland	19,877	60.3	1.90	Carmarthenshire	1,744	53.9	1.77
Trent	29,203	54.5	1.72				
				Swansea	2,449	54.6	1.73
WEST MIDLANDS	**65,956**	**61.1**	**1.91**	Neath Port Talbot	1,486	57.4	1.92
				Bridgend	1,523	59.5	1.97
Birmingham and the Black Country	32,167	66.9	2.04	Vale of Glamorgan	1,279	53.8	1.76
Shropshire and Staffordshire	16,639	57.2	1.86	Cardiff	3,955	51.8	1.56
West Midlands South	17,150	55.9	1.75				
				Rhondda Cynon Taff	2,864	60.7	1.92
EAST	**64,687**	**59.4**	**1.85**	Merthyr Tydfil	643	58.2	1.97
				Caerphilly	2,055	59.9	1.91
Bedfordshire and Hertfordshire	21,052	62.3	1.91	Blaenau Gwent	734	54.1	1.84
Essex	19,064	59.5	1.86	Torfaen	1,094	62.4	2.07
Norfolk, Suffolk and Cambridgeshire	24,571	57.1	1.78				
				Monmouthshire	819	53.4	1.84
LONDON	**116,019**	**63.8**	**1.79**	Newport	1,698	59.8	1.96
North Central London	18,935	61.2	1.72	**Normal residence outside**			
North East London	26,785	73.1	2.08	**England and Wales**	**214**	:	:
North West London	27,162	59.6	1.65				
South East London	24,049	66.1	1.90				
South West London	19,088	58.8	1.62				

Note: Rates are provisional - at subnational level, they are based on the most up-to-date mid-2004 population estimates released in December 2005. At national level, however, they are based on the 2004-based population projections for 2005.

1 Strategic Health Authorities in England and Local Health Boards in Wales.

2 The general fertility rate (GFR) is the number of live births per 1,000 women aged 15–44.

3 The total fertility rate (TFR) is the average number of children that would be born per woman if women experienced the age-specific fertility rates of the year in question throughout their childbearing lifespan.

Report:

Death registrations in England and Wales, 2005: area of residence

This report presents the numbers of deaths from all causes registered in England and Wales in 2005 and standardised mortality ratios (SMRs), both shown by area of usual residence of the deceased. Table 1 presents the data by administrative areas and Table 2 by health areas.

KEY OBSERVATIONS

- The highest level of mortality among the Government Office Regions in England in 2005, as represented by the SMR, was in the North East (113), followed by the North West (110). The lowest SMRs were in the South West (92) and South East (93).

- Among local authorities, the highest SMRs were in Hartlepool (131) then Knowsley (126), followed by Liverpool and Halton (both 125). The lowest occurred in Kensington and Chelsea (66), followed by Purbeck (77) and East Dorset (78).

- The local authorities with the highest and lowest SMRs were the same for both males and females. The highest SMR was in Hartlepool for males (130) and females (131), while the lowest occurred in the Kensington and Chelsea (62 for males and 69 for females).

- Among the Strategic Health Authorities in England, the highest SMRs occurred in County Durham and Tees Valley and Greater Manchester (both 115). The lowest occurred in North West London (88).

- In Wales, the highest SMRs were in both Merthyr Tydfil and Blaenau Gwent Local Health Boards (123), while the lowest SMR was in Monmouthshire (83).

EXPLANATORY NOTES

Occurrences and registrations

The year in which a death is registered may not correspond to the year in which the death occurred. Up to 1992 ONS publications gave numbers of deaths registered in the data year. However, since 1993 most of our published figures represent the number of deaths that occurred in the data

year. In most years (and for most causes of death) this change has little effect on annual totals. However, figures based on date of occurrence provide a more reliable basis for assessing the impact on mortality of external factors (such as 'flu outbreaks or cold weather), while registrations are more timely. We therefore, take two annual extracts from our deaths database.[1]

- The first annual extract, produced in April following the data year, comprises deaths that were registered in that year. Outputs produced using this extract include this report and a report by cause of death in the summer edition of *Health Statistics Quarterly*,[2] as well as the annual Vital Statistics tables.

- The second extract, produced in the September following the data year, comprises deaths that occurred in that year. This extract forms the basis for the mortality annual reference volumes in the DH series.

Standardised mortality ratios

To make meaningful comparisons of the level of mortality between different areas, it is necessary to take into account differences in their population structure. In Tables 1 and 2 this is done by using standardised mortality ratios (SMRs). These ratios, expressed in percentage terms, compare mortality in one population with mortality in a 'standard' population, while allowing for differences in age structure. For each area, the ratio is derived by comparing the number of deaths actually registered with the number that would have been expected if the mortality rates by sex and age for England and Wales applied to the area's population. If local mortality rates are high compared with national rates, the number of deaths observed will be greater than the expected number and the SMR will be greater than 100. However, for areas with low mortality SMRs will be less than 100. More details can be found in ONS annual reference volumes.[3]

As noted above the SMRs presented here allow for comparisons to be made with a national average as the results take into account differing age structures in the populations of local areas. Direct comparisons between areas or between the sexes can be misleading however, as can comparisons across time.

The SMRs allow the mortality experience of each local area to be compared to the national average. For example Rochdale local authority in the North West has an SMR for all persons of 120. This means that more deaths were observed than were expected in this area and mortality for this local authority is worse than for England and Wales as a whole. Wolverhampton local authority in the West Midlands, with an SMR for all persons of 109, also has a level of mortality that was worse than the national average. However, the lower SMR in Wolverhampton than in Rochdale does not necessarily imply that its mortality rates by age are lower. This is because the expected numbers of deaths are calculated by applying age-specific rates for England and Wales to the age distribution of each area. Consequently where two areas have identical death rates in every age group, but different population age structures, their SMRs may differ.

As well as presenting results for all persons, separate figures are also given for males and females separately. The latter were calculated using national age-specific death rates which were particular to each sex and each year. For this reason it is not possible to directly compare results for males and females and for different years.

Note that in this report the SMRs are provisional because they are based on the 2004 mid-year population estimates.

Population estimates

The population estimates used in this report are the most up-to-date at the time of its publication. Population estimates for mid-2004 were published on 20 December 2005 and included the correction made to females in the London Borough of Harrow. The estimates incorporate the findings of the local authority population studies, the results of which were published in July 2004. Further information on population estimates can be found on the National Statistics website (www.statistics.gov.uk/popest).

REFERENCES

1. Office for National Statistics (2005). *Mortality Statistics*: cause 2004, series DH2 no. 31, section 2.2.
2. Office for National Statistics (2006). Report: Death registrations in England and Wales, 2005: causes. *Health Statistics Quarterly* 30.
3. Office for National Statistics (2004). *Mortality Statistics: general 2003*, series DH1 no. 36, section 2.6.

Table 1	Deaths by local authority of usual residence, numbers and standardised mortality ratios (SMRs) by sex, 2005 registrations

England and Wales, Government Office Regions (within England), unitary authorities/counties/districts & London boroughs

Area of usual residence	Number of deaths			Standardised mortality ratios		
	Persons	Males	Females	Persons	Males	Females
ENGLAND AND WALES	**512,993**	**243,870**	**269,123**	**100**	**100**	**100**
ENGLAND	**479,678**	**227,956**	**251,722**	**99**	**99**	**100**
NORTH EAST	**27,449**	**12,951**	**14,498**	**113**	**112**	**113**
Darlington UA	1,077	485	592	107	105	109
Hartlepool UA	1,052	504	548	131	130	131
Middlesbrough UA	1,405	693	712	119	121	117
Redcar and Cleveland UA	1,481	709	772	107	108	106
Stockton-on-Tees UA	1,746	838	908	111	110	113
Durham	**5,557**	**2,580**	**2,977**	**116**	**114**	**118**
Chester-le-Street	563	253	310	116	109	123
Derwentside	1,014	481	533	115	117	113
Durham	838	404	434	111	112	110
Easington	1,094	509	585	123	120	125
Sedgefield	977	456	521	118	114	121
Teesdale	309	136	173	109	103	115
Wear Valley	762	341	421	117	115	118
Northumberland	**3,396**	**1,570**	**1,826**	**104**	**100**	**107**
Alnwick	376	178	198	100	100	101
Berwick-upon-Tweed	298	133	165	84	80	87
Blyth Valley	836	396	440	116	115	118
Castle Morpeth	539	245	294	97	87	107
Tynedale	639	298	341	97	96	99
Wansbeck	708	320	388	115	109	121
Tyne and Wear (Met County)	**11,735**	**5,572**	**6,163**	**113**	**114**	**112**
Gateshead	2,168	1,003	1,165	116	113	118
Newcastle upon Tyne	2,778	1,303	1,475	111	111	111
North Tyneside	2,080	985	1,095	104	106	102
South Tyneside	1,827	897	930	118	123	114
Sunderland	2,882	1,384	1,498	117	117	117
NORTH WEST	**71,210**	**33,926**	**37,284**	**110**	**112**	**109**
Blackburn with Darwen UA	1,318	632	686	121	124	119
Blackpool UA	1,976	963	1,013	121	127	115
Halton UA	1,170	567	603	125	123	127
Warrington UA	1,765	878	887	106	109	103
Cheshire	**6,722**	**3,145**	**3,577**	**97**	**95**	**98**
Chester	1,188	552	636	95	92	97
Congleton	847	400	447	92	91	93
Crewe and Nantwich	1,146	548	598	103	103	103
Ellesmere Port & Neston	747	353	394	96	94	98
Macclesfield	1,540	713	827	91	89	92
Vale Royal	1,254	579	675	105	101	109
Cumbria	**5,449**	**2,681**	**2,768**	**100**	**105**	**96**
Allerdale	1,097	542	555	106	110	103
Barrow-in-Furness	765	373	392	103	111	97
Carlisle	1,145	564	581	106	111	102
Copeland	747	391	356	111	120	103
Eden	547	273	274	95	97	93
South Lakeland	1,148	538	610	85	87	84
Greater Manchester (Met County)	**25,539**	**12,257**	**13,282**	**115**	**117**	**112**
Bolton	2,790	1,333	1,457	119	120	118
Bury	1,730	795	935	107	106	108
Manchester	3,993	2,001	1,992	121	129	114
Oldham	2,236	1,075	1,161	119	123	115
Rochdale	2,099	998	1,101	120	120	119
Salford	2,382	1,158	1,224	119	125	114
Stockport	2,888	1,379	1,509	102	106	99
Tameside	2,332	1,080	1,252	121	120	121
Trafford	2,074	999	1,075	100	101	99
Wigan	3,015	1,439	1,576	117	118	116

Note: SMRs are based on mid-2004 population estimates with 2005 live births (used for calculations involving deaths under 1 year).

<table>
<tr><td>Table 1
continued</td><td colspan="6">Deaths by local authority of usual residence, numbers and standardised mortality ratios (SMRs) by sex, 2005 registrations</td></tr>
</table>

England and Wales, Government Office Regions (within England), unitary authorities/counties/districts & London boroughs

Area of usual residence	Number of deaths			Standardised mortality ratios		
	Persons	Males	Females	Persons	Males	Females
Lancashire	**12,278**	**5,719**	**6,559**	**108**	**107**	**109**
Burnley	911	425	486	111	108	113
Chorley	982	468	514	111	111	111
Fylde	1,011	431	580	98	91	104
Hyndburn	834	401	433	112	115	109
Lancaster	1,469	676	793	101	101	100
Pendle	927	446	481	111	114	108
Preston	1,329	648	681	121	121	121
Ribble Valley	546	245	301	95	91	98
Rossendale	690	313	377	121	117	124
South Ribble	919	441	478	93	94	93
West Lancashire	1,131	521	610	111	107	114
Wyre	1,529	704	825	111	109	112
Merseyside (Met County)	**14,993**	**7,084**	**7,909**	**115**	**117**	**113**
Knowsley	1,504	738	766	126	130	123
Liverpool	4,658	2,245	2,413	125	127	123
Sefton	3,338	1,522	1,816	107	108	107
St Helens	1,889	919	970	119	124	114
Wirral	3,604	1,660	1,944	105	107	104
YORKSHIRE AND THE HUMBER	**50,564**	**23,990**	**26,574**	**104**	**105**	**104**
East Riding of Yorkshire UA	3,489	1,642	1,847	97	96	98
Kingston upon Hull, City of UA	2,577	1,258	1,319	120	122	117
North East Lincolnshire UA	1,631	805	826	104	110	99
North Lincolnshire UA	1,613	798	815	103	107	101
York UA	1,694	853	841	91	97	84
North Yorkshire	**5,970**	**2,811**	**3,159**	**91**	**91**	**92**
Craven	587	265	322	85	82	87
Hambleton	800	415	385	88	92	84
Harrogate	1,474	650	824	87	84	90
Richmondshire	460	201	259	99	90	108
Ryedale	582	307	275	90	97	84
Scarborough	1,399	648	751	99	100	98
Selby	668	325	343	94	94	93
South Yorkshire (Met County)	**13,511**	**6,336**	**7,175**	**111**	**110**	**111**
Barnsley	2,469	1,171	1,298	119	120	118
Doncaster	3,077	1,482	1,595	113	112	113
Rotherham	2,622	1,224	1,398	112	110	114
Sheffield	5,343	2,459	2,884	105	105	106
West Yorkshire (Met County)	**20,079**	**9,487**	**10,592**	**106**	**107**	**105**
Bradford	4,537	2,083	2,454	111	110	113
Calderdale	1,943	894	1,049	105	105	105
Kirklees	3,754	1,738	2,016	107	106	108
Leeds	6,583	3,210	3,373	100	103	97
Wakefield	3,262	1,562	1,700	111	112	110
EAST MIDLANDS	**42,007**	**20,153**	**21,854**	**101**	**100**	**102**
Derby UA	2,210	1,117	1,093	99	103	95
Leicester UA	2,649	1,271	1,378	115	116	114
Nottingham UA	2,510	1,280	1,230	113	118	107
Rutland UA	300	146	154	79	80	79
Derbyshire	**7,872**	**3,693**	**4,179**	**103**	**101**	**105**
Amber Valley	1,196	570	626	98	100	96
Bolsover	887	414	473	118	116	120
Chesterfield	1,182	542	640	112	110	114
Derbyshire Dales	739	320	419	89	82	95
Erewash	1,074	536	538	100	104	95
High Peak	911	433	478	102	100	103
North East Derbyshire	1,111	496	615	105	96	115
South Derbyshire	772	382	390	103	102	104
Leicestershire	**5,803**	**2,713**	**3,090**	**97**	**92**	**102**
Blaby	720	337	383	85	79	91
Charnwood	1,442	696	746	101	101	102

<table>
<tr><td>Table I continued</td><td>Deaths by local authority of usual residence, numbers and standardised mortality ratios (SMRs) by sex, 2005 registrations</td></tr>
</table>

England and Wales, Government Office Regions (within England), unitary authorities/counties/districts & London boroughs

Area of usual residence	Number of deaths			Standardised mortality ratios		
	Persons	Males	Females	Persons	Males	Females
Harborough	746	373	373	96	94	98
Hinckley and Bosworth	986	418	568	98	86	110
Melton	465	212	253	94	89	99
North West Leicestershire	907	424	483	106	102	110
Oadby and Wigston	537	253	284	94	90	97
Lincolnshire	**7,306**	**3,543**	**3,763**	**97**	**95**	**98**
Boston	685	330	355	101	101	101
East Lindsey	1,681	823	858	95	93	98
Lincoln	849	405	444	104	106	103
North Kesteven	963	453	510	89	84	94
South Holland	917	469	448	92	95	90
South Kesteven	1,253	608	645	95	95	95
West Lindsey	958	455	503	105	100	109
Northamptonshire	**5,671**	**2,720**	**2,951**	**99**	**98**	**101**
Corby	507	256	251	119	118	120
Daventry	584	289	295	92	92	93
East Northamptonshire	814	359	455	107	99	114
Kettering	849	392	457	105	103	106
Northampton	1,650	792	858	99	98	100
South Northamptonshire	634	314	320	87	85	89
Wellingborough	633	318	315	94	95	93
Nottinghamshire	**7,686**	**3,670**	**4,016**	**102**	**101**	**102**
Ashfield	1,207	598	609	113	118	109
Bassetlaw	1,141	540	601	106	103	109
Broxtowe	1,040	500	540	97	96	97
Gedling	1,091	517	574	95	93	96
Mansfield	1,093	540	553	114	116	111
Newark and Sherwood	1,151	542	609	101	99	102
Rushcliffe	963	433	530	89	84	94
WEST MIDLANDS	**52,980**	**25,743**	**27,237**	**104**	**105**	**102**
Herefordshire, County of UA	**2,009**	**965**	**1,044**	**95**	**95**	**96**
Stoke-on-Trent UA	**2,572**	**1,260**	**1,312**	**114**	**119**	**109**
Telford and Wrekin UA	**1,309**	**664**	**645**	**104**	**108**	**100**
Shropshire	**3,006**	**1,449**	**1,557**	**94**	**94**	**93**
Bridgnorth	591	306	285	105	111	99
North Shropshire	602	279	323	94	91	97
Oswestry	407	201	206	97	100	94
Shrewsbury and Atcham	956	448	508	92	93	92
South Shropshire	450	215	235	82	80	84
Staffordshire	**8,174**	**3,934**	**4,240**	**105**	**105**	**105**
Cannock Chase	919	461	458	119	123	115
East Staffordshire	1,022	491	531	104	104	103
Lichfield	927	422	505	100	94	106
Newcastle-under-Lyme	1,311	637	674	105	108	103
South Staffordshire	1,140	573	567	107	111	104
Stafford	1,305	620	685	101	101	102
Staffordshire Moorlands	1,008	490	518	101	100	102
Tamworth	542	240	302	103	93	113
Warwickshire	**5,212**	**2,460**	**2,752**	**101**	**99**	**103**
North Warwickshire	598	292	306	105	106	105
Nuneaton and Bedworth	1,214	589	625	113	113	113
Rugby	953	456	497	106	105	107
Stratford-on-Avon	1,224	534	690	94	87	101
Warwick	1,223	589	634	93	94	92
West Midlands (Met County)	**25,117**	**12,355**	**12,762**	**106**	**109**	**103**
Birmingham	9,129	4,507	4,622	108	113	103
Coventry	2,987	1,473	1,514	107	109	104
Dudley	3,129	1,530	1,599	105	107	103
Sandwell	3,085	1,503	1,582	115	119	111
Solihull	1,762	866	896	87	89	85

<table>
<tr><td>Table 1
continued</td><td>Deaths by local authority of usual residence, numbers and standardised mortality ratios (SMRs) by sex, 2005 registrations</td></tr>
</table>

England and Wales, Government Office Regions (within England), unitary authorities/counties/districts & London boroughs

Area of usual residence	Number of deaths			Standardised mortality ratios		
	Persons	Males	Females	Persons	Males	Females
Walsall	2,442	1,222	1,220	102	107	98
Wolverhampton	2,583	1,254	1,329	109	109	109
Worcestershire	**5,581**	**2,656**	**2,925**	**98**	**98**	**99**
Bromsgrove	970	462	508	100	100	101
Malvern Hills	943	426	517	98	93	103
Redditch	685	338	347	107	107	107
Worcester	793	378	415	96	98	95
Wychavon	1,182	593	589	94	97	92
Wyre Forest	1,008	459	549	98	95	101
EAST	**52,499**	**24,880**	**27,619**	**95**	**93**	**96**
Luton UA	**1,464**	**731**	**733**	**110**	**107**	**113**
Peterborough UA	**1,488**	**711**	**777**	**111**	**109**	**113**
Southend-on-Sea UA	**1,956**	**859**	**1,097**	**101**	**101**	**101**
Thurrock UA	**1,132**	**538**	**594**	**99**	**99**	**99**
Bedfordshire	**3,340**	**1,633**	**1,707**	**98**	**98**	**99**
Bedford	1,345	621	724	97	92	101
Mid Bedfordshire	984	494	490	96	95	96
South Bedfordshire	1,011	518	493	104	108	99
Cambridgeshire	**4,849**	**2,313**	**2,536**	**91**	**89**	**93**
Cambridge	858	391	467	91	89	93
East Cambridgeshire	641	324	317	84	86	82
Fenland	1,063	499	564	108	105	111
Huntingdonshire	1,232	551	681	91	82	99
South Cambridgeshire	1,055	548	507	83	87	80
Essex	**12,911**	**6,121**	**6,790**	**94**	**94**	**95**
Basildon	1,459	697	762	100	101	100
Braintree	1,267	565	702	97	92	101
Brentwood	648	289	359	84	79	88
Castle Point	831	386	445	91	86	96
Chelmsford	1,235	591	644	85	85	86
Colchester	1,392	679	713	94	97	90
Epping Forest	1,247	588	659	99	98	101
Harlow	651	323	328	100	102	97
Maldon	563	279	284	94	96	91
Rochford	730	371	359	87	90	83
Tendring	2,273	1,075	1,198	100	101	99
Uttlesford	615	278	337	89	82	95
Hertfordshire	**9,032**	**4,134**	**4,898**	**93**	**89**	**96**
Broxbourne	666	306	360	89	83	94
Dacorum	1,225	543	682	95	87	102
East Hertfordshire	994	455	539	88	84	92
Hertsmere	996	431	565	103	94	110
North Hertfordshire	1,181	560	621	98	99	97
St Albans	1,022	479	543	86	84	88
Stevenage	622	307	315	97	98	95
Three Rivers	798	340	458	91	83	98
Watford	669	292	377	102	95	108
Welwyn Hatfield	859	421	438	85	87	84
Norfolk	**9,240**	**4,588**	**4,652**	**93**	**95**	**91**
Breckland	1,410	688	722	95	95	96
Broadland	1,404	657	747	98	94	103
Great Yarmouth	1,143	574	569	103	108	98
King's Lynn and West Norfolk	1,629	819	810	94	95	92
North Norfolk	1,380	706	674	90	94	86
Norwich	1,165	601	564	92	104	82
South Norfolk	1,109	543	566	83	81	84
Suffolk	**7,087**	**3,252**	**3,835**	**92**	**87**	**96**
Babergh	876	396	480	89	83	95
Forest Heath	475	221	254	92	86	97
Ipswich	1,124	512	612	95	91	98
Mid Suffolk	816	384	432	85	80	89
St Edmundsbury	998	464	534	96	92	101

<table>
<tr><td>Table 1
continued</td><td colspan="6">Deaths by local authority of usual residence, numbers and standardised mortality ratios (SMRs) by sex, 2005 registrations</td></tr>
</table>

England and Wales, Government Office Regions (within England), unitary authorities/counties/districts & London boroughs

Area of usual residence	Number of deaths			Standardised mortality ratios		
	Persons	Males	Females	Persons	Males	Females
Suffolk Coastal	1,368	628	740	89	86	92
Waveney	1,430	647	783	94	89	99
LONDON	**52,992**	**25,883**	**27,109**	**95**	**97**	**94**
Inner London	**17,693**	**9,069**	**8,624**	**98**	**101**	**96**
Camden	1,344	748	596	100	109	90
City of London[1]	38	20	18	:	:	:
Hackney	1,155	616	539	98	106	90
Hammersmith and Fulham	982	504	478	89	94	84
Haringey	1,205	657	548	94	106	84
Islington	1,146	597	549	108	111	106
Kensington and Chelsea	910	431	479	66	62	69
Lambeth	1,624	815	809	108	107	110
Lewisham	1,874	902	972	112	113	111
Newham	1,541	775	766	115	112	117
Southwark	1,599	832	767	100	106	94
Tower Hamlets	1,141	625	516	107	110	104
Wandsworth	1,869	911	958	103	107	100
Westminster	1,265	636	629	80	79	82
Outer London	**35,299**	**16,814**	**18,485**	**94**	**94**	**93**
Barking and Dagenham	1,536	727	809	111	113	110
Barnet	2,569	1,162	1,407	87	86	89
Bexley	1,895	903	992	91	90	91
Brent	1,638	864	774	89	94	85
Bromley	2,672	1,230	1,442	86	86	86
Croydon	2,525	1,183	1,342	94	89	99
Ealing	2,002	1,030	972	93	96	91
Enfield	2,208	1,004	1,204	94	91	96
Greenwich	1,952	983	969	112	124	102
Harrow	1,616	767	849	85	85	85
Havering	2,337	1,065	1,272	102	98	105
Hillingdon	1,929	927	1,002	93	95	92
Hounslow	1,538	769	769	105	106	105
Kingston upon Thames	1,156	542	614	91	90	91
Merton	1,304	606	698	85	85	86
Redbridge	2,008	967	1,041	97	98	95
Richmond upon Thames	1,256	573	683	81	78	83
Sutton	1,497	702	795	92	94	90
Waltham Forest	1,661	810	851	106	111	100
SOUTH EAST	**76,849**	**35,399**	**41,450**	**93**	**91**	**94**
Bracknell Forest UA	742	359	383	95	96	94
Brighton and Hove UA	2,425	1,121	1,304	96	98	94
Isle of Wight UA	1,664	769	895	89	89	88
Medway UA	2,101	1,010	1,091	108	108	107
Milton Keynes UA	1,506	709	797	105	101	109
Portsmouth UA	1,704	811	893	96	98	94
Reading UA	1,097	518	579	97	97	96
Slough UA	779	367	412	92	85	100
Southampton UA	1,957	955	1,002	99	102	96
West Berkshire UA	1,090	528	562	89	88	89
Windsor and Maidenhead UA	1,147	536	611	91	89	92
Wokingham UA	981	467	514	84	80	87
Buckinghamshire	**4,055**	**1,948**	**2,107**	**92**	**91**	**92**
Aylesbury Vale	1,411	674	737	102	99	104
Chiltern	783	371	412	84	84	84
South Bucks	630	272	358	93	85	100
Wycombe	1,231	631	600	86	91	82
East Sussex	**6,389**	**2,835**	**3,554**	**89**	**87**	**91**
Eastbourne	1,367	574	793	94	91	96
Hastings	977	424	553	101	100	102
Lewes	1,105	511	594	82	82	83
Rother	1,373	613	760	89	89	90
Wealden	1,567	713	854	83	81	86

1 SMRs for City of London and Isles of Scilly have not been calculated because of the very small numbers of deaths and populations in these areas.

<table>
<tr><td>Table 1
continued</td><td colspan="6">Deaths by local authority of usual residence, numbers and standardised mortality ratios (SMRs) by sex, 2005 registrations</td></tr>
</table>

England and Wales, Government Office Regions (within England), unitary authorities/counties/districts & London boroughs

Area of usual residence	Number of deaths			Standardised mortality ratios		
	Persons	Males	Females	Persons	Males	Females
Hampshire	**11,353**	**5,332**	**6,021**	**89**	**87**	**90**
Basingstoke and Deane	1,062	487	575	87	82	92
East Hampshire	1,045	456	589	92	85	99
Eastleigh	990	439	551	91	84	97
Fareham	983	475	508	84	84	84
Gosport	815	389	426	106	108	104
Hart	587	291	296	83	84	83
Havant	1,216	582	634	92	91	94
New Forest	2,031	970	1,061	83	83	82
Rushmoor	630	306	324	94	95	93
Test Valley	997	477	520	93	95	91
Winchester	997	460	537	83	81	84
Kent	**13,964**	**6,443**	**7,521**	**98**	**96**	**100**
Ashford	961	447	514	91	86	96
Canterbury	1,610	714	896	97	95	100
Dartford	850	398	452	114	111	117
Dover	1,242	574	668	101	100	102
Gravesham	849	420	429	98	102	94
Maidstone	1,379	654	725	100	99	101
Sevenoaks	939	433	506	79	77	81
Shepway	1,244	552	692	102	98	106
Swale	1,139	528	611	100	95	104
Thanet	1,867	882	985	110	115	107
Tonbridge and Malling	920	413	507	92	85	98
Tunbridge Wells	964	428	536	90	88	92
Oxfordshire	**5,111**	**2,370**	**2,741**	**90**	**87**	**93**
Cherwell	1,004	479	525	87	87	87
Oxford	994	462	532	89	87	91
South Oxfordshire	1,142	532	610	90	88	92
Vale of White Horse	1,018	470	548	89	84	95
West Oxfordshire	953	427	526	94	89	99
Surrey	**9,824**	**4,432**	**5,392**	**89**	**86**	**92**
Elmbridge	1,118	463	655	85	76	93
Epsom and Ewell	598	299	299	82	89	76
Guildford	999	448	551	81	77	84
Mole Valley	852	384	468	88	84	91
Reigate and Banstead	1,317	603	714	100	98	101
Runnymede	722	322	400	89	86	92
Spelthorne	886	434	452	98	98	98
Surrey Heath	653	284	369	94	86	101
Tandridge	796	366	430	91	87	95
Waverley	1,197	533	664	88	86	90
Woking	686	296	390	83	76	89
West Sussex	**8,960**	**3,889**	**5,071**	**92**	**87**	**96**
Adur	774	343	431	98	96	100
Arun	2,215	986	1,229	94	94	94
Chichester	1,322	595	727	87	84	90
Crawley	788	368	420	90	79	103
Horsham	1,073	479	594	81	76	85
Mid Sussex	1,255	513	742	93	83	102
Worthing	1,533	605	928	103	98	106
SOUTH WEST	**53,128**	**25,031**	**28,097**	**92**	**91**	**92**
Bath and North East Somerset UA	1,698	778	920	90	87	93
Bournemouth UA	1,996	873	1,123	92	91	94
Bristol, City of UA	3,559	1,680	1,879	101	102	101
North Somerset UA	2,193	1,005	1,188	93	91	95
Plymouth UA	2,390	1,148	1,242	101	105	97
Poole UA	1,579	763	816	89	92	86
South Gloucestershire UA	1,926	964	962	87	89	86
Swindon UA	1,583	776	807	105	105	106
Torbay UA	1,820	861	959	96	101	91
Cornwall and Isles of Scilly	**5,917**	**2,816**	**3,101**	**93**	**93**	**93**
Caradon	908	440	468	94	95	93
Carrick	1,094	532	562	92	96	88
Kerrier	1,070	523	547	97	99	95

Table 1 continued	**Deaths by local authority of usual residence, numbers and standardised mortality ratios (SMRs) by sex, 2005 registrations**

England and Wales, Government Office Regions (within England), unitary authorities/counties/districts & London boroughs

Area of usual residence	Number of deaths			Standardised mortality ratios		
	Persons	Males	Females	Persons	Males	Females
North Cornwall	1,005	471	534	96	94	97
Penwith	793	364	429	93	93	94
Restormel	1,040	482	558	90	86	92
Isles of Scilly[1]	7	4	3	:	:	:
Devon	**8,268**	**3,925**	**4,343**	**88**	**90**	**87**
East Devon	1,916	884	1,032	89	88	90
Exeter	1,053	500	553	93	100	87
Mid Devon	743	371	372	90	94	87
North Devon	1,001	448	553	90	86	95
South Hams	917	453	464	86	89	84
Teignbridge	1,412	693	719	84	88	80
Torridge	707	338	369	91	90	92
West Devon	519	238	281	84	82	85
Dorset	**4,790**	**2,295**	**2,495**	**84**	**83**	**84**
Christchurch	634	299	335	79	78	80
East Dorset	1,030	506	524	78	77	79
North Dorset	683	337	346	86	87	86
Purbeck	461	230	231	77	79	75
West Dorset	1,254	586	668	88	87	88
Weymouth and Portland	728	337	391	92	91	92
Gloucestershire	**5,671**	**2,615**	**3,056**	**92**	**89**	**94**
Cheltenham	1,059	463	596	88	83	91
Cotswold	894	392	502	88	82	94
Forest of Dean	776	381	395	90	93	88
Gloucester	1,026	481	545	101	97	104
Stroud	1,137	541	596	93	94	92
Tewkesbury	779	357	422	89	86	93
Somerset	**5,632**	**2,601**	**3,031**	**92**	**90**	**94**
Mendip	1,126	515	611	97	95	99
Sedgemoor	1,173	555	618	93	93	94
South Somerset	1,674	775	899	90	88	92
Taunton Deane	1,191	546	645	94	95	94
West Somerset	468	210	258	80	75	85
Wiltshire	4,106	1,931	2,175	90	89	92
Kennet	718	337	381	94	93	96
North Wiltshire	1,088	529	559	93	92	94
Salisbury	1,146	521	625	88	85	91
West Wiltshire	1,154	544	610	87	86	89
WALES	**32,162**	**15,209**	**16,953**	**105**	**105**	**105**
Isle of Anglesey	805	385	420	102	103	102
Gwynedd	1,307	616	691	98	100	96
Conwy	1,509	692	817	97	97	98
Denbighshire	1,220	556	664	102	102	102
Flintshire	1,443	656	787	105	99	110
Wrexham	1,373	647	726	108	108	107
Powys	1,515	736	779	96	93	98
Ceredigion	766	381	385	86	88	85
Pembrokeshire	1,458	693	765	109	109	109
Carmarthenshire	2,168	1,039	1,129	106	108	104
Swansea	2,555	1,213	1,342	104	104	104
Neath Port Talbot	1,544	720	824	105	106	104
Bridgend	1,460	721	739	114	117	111
The Vale of Glamorgan	1,251	586	665	100	97	103
Cardiff	2,745	1,313	1,432	101	102	101
Rhondda, Cynon, Taff	2,602	1,244	1,358	117	120	114
Merthyr Tydfil	637	301	336	123	125	121
Caerphilly	1,768	805	963	115	110	119
Blaenau Gwent	859	392	467	123	122	124
Torfaen	951	448	503	104	104	104
Monmouthshire	806	398	408	83	86	81
Newport	1,420	667	753	106	106	106
Normal residence outside England and Wales	**1,153**	**705**	**448**	..	..	..

1 SMRs for City of London and Isles of Scilly have not been calculated because of the very small numbers of deaths and populations in these areas.

Table 2

Deaths by health area of usual residence, numbers and standardised mortality ratios (SMRs) by sex, 2005 registrations

England and Wales, Government Office Regions (within England) and health authorities/boards[1]

Area of usual residence	Number of deaths			Standardised mortality ratios		
	Persons	Males	Females	Persons	Males	Females
ENGLAND AND WALES	**512,993**	**243,870**	**269,123**	**100**	**100**	**100**
ENGLAND	**479,678**	**227,956**	**251,722**	**99**	**99**	**100**
NORTH EAST	**27,449**	**12,951**	**14,498**	**113**	**112**	**113**
County Durham and Tees Valley	12,318	5,809	6,509	115	114	116
Northumberland, Tyne & Wear	15,131	7,142	7,989	111	110	111
NORTH WEST	**71,210**	**33,926**	**37,284**	**110**	**112**	**109**
Cheshire & Merseyside	24,650	11,674	12,976	109	110	108
Cumbria and Lancashire	21,021	9,995	11,026	107	109	106
Greater Manchester	25,539	12,257	13,282	115	117	112
YORKSHIRE AND THE HUMBER	**50,564**	**23,990**	**26,574**	**104**	**105**	**104**
North and East Yorkshire and Northern Lincolnshire	16,974	8,167	8,807	98	100	97
South Yorkshire	13,511	6,336	7,175	111	110	111
West Yorkshire	20,079	9,487	10,592	106	107	105
EAST MIDLANDS	**42,007**	**20,153**	**21,854**	**101**	**100**	**102**
Leicestershire, Northamptonshire and Rutland	14,423	6,850	7,573	100	98	103
Trent	27,584	13,303	14,281	101	101	102
WEST MIDLANDS	**52,980**	**25,743**	**27,237**	**104**	**105**	**102**
Birmingham and the Black Country	22,130	10,882	11,248	106	109	103
Shropshire and Staffordshire	15,061	7,307	7,754	104	105	103
West Midlands South[2]	15,789	7,554	8,235	100	100	101
EAST	**52,499**	**24,880**	**27,619**	**95**	**93**	**96**
Bedfordshire and Hertfordshire	13,836	6,498	7,338	96	93	99
Essex	15,999	7,518	8,481	95	95	96
Norfolk, Suffolk and Cambridgeshire	22,664	10,864	11,800	93	92	94
LONDON	**52,992**	**25,883**	**27,109**	**95**	**97**	**94**
North Central London	8,472	4,168	4,304	94	97	92
North East London	11,417	5,605	5,812	104	105	103
North West London	11,880	5,928	5,952	88	89	87
South East London	11,616	5,665	5,951	99	102	97
South West London	9,607	4,517	5,090	92	91	92
SOUTH EAST	**76,849**	**35,399**	**41,450**	**93**	**91**	**94**
Hampshire and Isle of Wight	16,678	7,867	8,811	90	90	91
Kent and Medway	16,065	7,453	8,612	99	98	101
Surrey and Sussex	27,598	12,277	15,321	91	88	93
Thames Valley	16,508	7,802	8,706	92	90	94
SOUTH WEST	**53,128**	**25,031**	**28,097**	**92**	**91**	**92**
Avon, Gloucestershire and Wiltshire	20,736	9,749	10,987	93	92	95
Dorset and Somerset	13,997	6,532	7,465	89	88	89
South West Peninsula	18,395	8,750	9,645	92	94	91
WALES	**32,162**	**15,209**	**16,953**	**105**	**105**	**105**
Anglesey	805	385	420	102	103	102
Gwynedd	1,307	616	691	98	100	96
Conwy	1,509	692	817	97	97	98
Denbighshire	1,220	556	664	102	102	102
Flintshire	1,443	656	787	105	99	110

Note: SMRs are based on mid-2004 population estimates with 2005 live births (used for calculations involving deaths under 1 year).

1 Strategic Health Authorities in England and Local Health Boards in Wales.
2 Renamed West Midlands South on 9 February 2004 (formerly known as Coventry, Warwickshire, Herefordshire and Worcestershire).

Table 2 continued	Deaths by health area of usual residence, numbers and standardised mortality ratios (SMRs) by sex, 2005 registrations

England and Wales, Government Office Regions (within England) and health authorities/boards[1]

Area of usual residence	Number of deaths			Standardised mortality ratios		
	Persons	Males	Females	Persons	Males	Females
Wrexham	1,373	647	726	108	108	107
Powys	1,515	736	779	96	93	98
Ceredigion	766	381	385	86	88	85
Pembrokeshire	1,458	693	765	109	109	109
Carmarthenshire	2,168	1,039	1,129	106	108	104
Swansea	2,555	1,213	1,342	104	104	104
Neath Port Talbot	1,544	720	824	105	106	104
Bridgend	1,460	721	739	114	117	111
Vale of Glamorgan	1,251	586	665	100	97	103
Cardiff	2,745	1,313	1,432	101	102	101
Rhondda, Cynon, Taff	2,602	1,244	1,358	117	120	114
Merthyr Tydfil	637	301	336	123	125	121
Caerphilly	1,768	805	963	115	110	119
Blaenau Gwent	859	392	467	123	122	124
Torfaen	951	448	503	104	104	104
Monmouthshire	806	398	408	83	86	81
Newport	1,420	667	753	106	106	106
Normal residence outside England and Wales	**1,153**	**705**	**448**	..	..	..

1 Strategic Health Authorities in England and Local Health Boards in Wales.

Other population and health articles, publications and data

Health Statistics Quarterly 31
Publication 24 August 2006

Planned articles:
- Net cohort migration in England and Wales: Low past birth trends may influence net migration
- Population definitions and the future of statistical provision – what do users want
- Administrative sources and population statistics
- Estimates of the population by ethnic group in England

Reports:
- Live births in England and Wales, 2005: area of residence
- Death registrations in England and Wales, 2005: area of residence

Population Trends 125
Publication 28 September 2006

Planned articles:
- Estimation of the England and Wales population: a comparative study of the cohort component method
- Developments with ONS's Small Area Population Estimates project
- Profile of the UK by NS-SEC

Forthcoming Annual Reference Volumes

Title	Planned publication
Birth statistics 2005, FM1 no.34*	December 2006
Cancer statistics: registrations 2004, MB1 no.35*	December 2006
Congenital anomaly statistics 2005, MB3 no.20*	December 2006
Mortality statistics: cause 2005, DH2 no.32*	December 2006

* Available through the National Statistics website only; www.statistics.gov.uk